T5-AFM-643

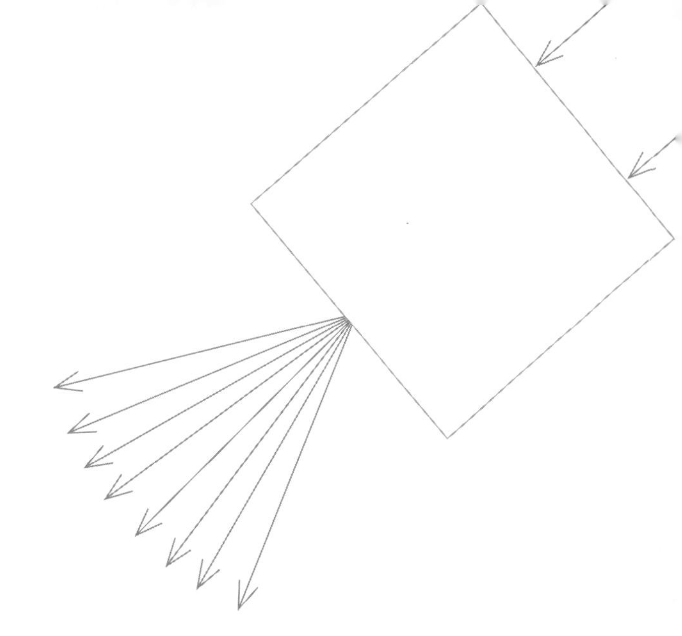

Markov-Modulated Processes & Semiregenerative Phenomena

António Pacheco
Technical University of Lisbon, Portugal

Loon Ching Tang
National University of Singapore, Singapore

Narahari U Prabhu
Cornell University, USA

Markov-Modulated Processes & Semiregenerative Phenomena

NEW JERSEY • LONDON • SINGAPORE • BEIJING • SHANGHAI • HONG KONG • TAIPEI • CHENNAI

Published by

World Scientific Publishing Co. Pte. Ltd.

5 Toh Tuck Link, Singapore 596224

USA office: 27 Warren Street, Suite 401-402, Hackensack, NJ 07601

UK office: 57 Shelton Street, Covent Garden, London WC2H 9HE

British Library Cataloguing-in-Publication Data
A catalogue record for this book is available from the British Library.

MARKOV-MODULATED PROCESSES AND SEMIREGENERATIVE PHENOMENA

ISBN-13 978-981-279-318-8
ISBN-10 981-279-318-6

Printed in Singapore by World Scientific Printers

Preface

Markov-modulated processes are processes that are modulated (or driven) by an underlying Markov process. The earliest of these processes studied in the literature are Markov-additive processes, namely, Lévy processes with a Markov component. Semiregenerative phenomena are regenerative phenomena that have, in addition, a Markov component. Such phenomena are important in the study of Markov-additive processes and arise in several important applications.

Our presentation emphasizes the interplay between theory and applications of stochastic processes. Accordingly, the first six chapters are devoted to theory along with illustrative examples, while the last three chapters treat applications to queueing, data communication, and storage systems. As a result, the book may be used both as a research monograph and as a textbook for a graduate course in applied stochastic processes focusing on Markov-modulated processes, semiregenerative phenomena, and their applications. Although the book is more easily read in a linear fashion, care was taken to give the reader the possibility of reading a chapter separately as proper referencing of results used from other chapters is done in a systematic manner.

The book starts with a brief introduction to recurrence and regeneration, which contains historical insights and brings out the connection between recurrent phenomena and renewal processes. This introduction, given in Chapter 1, prepares the reader for subsequent results in the book on semiregenerative phenomena, which are linked regenerative phenomena that have, in addition, a Markov property.

Following the introduction to recurrence and regeneration, we present in Chapter 2 an exposition of the basic concepts and main properties of Markov renewal processes (MRPs), along with the properties of some special cases of Markov-additive processes (MAPs) which are of importance in applied probability. Our approach makes it clear that an MRP is a family of Markov-dependent renewal processes, whose corresponding family of renewal counting processes gives rise in a natural fashion to the so-called Markov renewal equation. We establish the connection between the existence and uniqueness of the solution of this equation and the finiteness of the total number of renewals. Among the MAPs considered in the chapter is the so called MMPP (Markov-modulated Poisson process), in which the rate of occurrence of Poisson events changes instantaneously with the changes of state of the modulating Markov chain, along with other Markov-compound Poisson processes. We also illustrate the use of infinitesimal generators for MAPs.

The theory of semiregenerative processes is developed in Chapter 3, where we explore the correspondence between semiregenerative sets and the range of a Markov subordinator with a unit drift, in the continuous-time case, or an MRP in the discrete-time case. In the former case, we construct a Markov subordinator with a unit drift whose range turns out to be a semiregenerative set. In the case where the label parameter set is finite we prove the converse, that every semiregenerative set corresponds to the range of a Markov subordinator. In the latter case, we show that the semirecurrent set of a semirecurrent phenomena corresponds to the range of an MRP and, conversely, a semirecurrent set can only arise in this manner.

In Chapter 4 we present some basic theory of the Markov random walk (MRW) viewing it as a family of imbedded standard random walks and consider a time-reversed version of the MRW which plays a key role in developing results analogous to those of the classical random walk. One of the central themes of the chapter is on the fluctuation theory of MRWs, first studied by Newbould in the 1970's for nondegenerate MRWs with a finite state ergodic Markov chain. Here, more general results for MRWs with countable state space are presented and a fundamental definition for degenerate MRW is given so that the probabilistic structure of MRWs can be firmly established. Together with the main result containing Wiener-Hopf factorization of the underlying transition measures based on the associated semirecurrent sets, it provides better insights for models with Markov sources and facilitates potential applications.

In Chapter 5 we establish the central limit theorem and of iterated logarithm for the additive component of an MRW. Our proofs are probabilistic using i.i.d. sequences of random variables denoting normalized increments of the additive component of the MRW between successive visits of the modulating Markov chain to a given state. We express the parameters involved in terms of the transition measure of the MRW, and indicate an application of the theorems derived in the chapter to finance models.

In Chapter 6 we address MAPs of arrivals, and investigate their partial lack of memory property, interarrival times, moments of the number of counts, and limit theorems for the number of counts. We further consider transformations of MAPs of arrivals that preserve the Markov-additive property, such as linear transformations, patching of independent processes, linear combinations, and some random time transformations. Moreover, we consider secondary recordings that generate new arrival processes from an original MAP of arrivals; these include, in particular, marking, colouring and thinning. For Markov-Bernoulli recording we show that the secondary process in each case turns out to be an MAP of arrivals.

In Chapter 7 we consider single server queueing systems that are modulated by a discrete time Markov chain on a countable state space. The underlying stochastic process is an MRW whose increments can be expressed as differences between service times and interarrival times. We derive the joint distributions of the waiting and idle times in the presence of the modulating Markov chain. Our approach is based on properties of the ladder sets associated with this MRW and its time-reversed counterpart. The special case of a Markov-modulated M/M/1 queueing system is then analyzed and results analogous to the classical case are obtained.

In Chapter 8 we investigate a storage model whose input process (X, J) is an MAP, where, in addition to a nonnegative drift $a(j)$ when the modulating Markov chain J with countable state space is in state j, X has nonnegative jumps whose rate and size depend on the state transitions of J. This formulation provides for the possibility of two sources of input, one slow source bringing in data in a fluid fashion and the other bringing in packets. As for the demand for transmission of data we assume that it arises at a rate $d(j)$ when the current state of J is j, and the storage policy is to meet the demand if physically possible. We prove the existence and uniqueness of the solution of the basic integral equation associated to the model and investigate various processes of interest, showing in particular that the busy-period process is a Markov-compound Poisson process. Transforms of the various processes of interest are obtained and the steady state behavior of the model is investigated.

Finally, in Chapter 9 we investigate a storage model where the input and the demand are continuous additive functionals on a modulating Markov chain, with the instantaneous input and demand rates being only a function of the state of that chain. As in the model considered in Chapter 8, the storage policy is to meet the largest possible portion of the demand. Our analysis is based on the net input process imbedded at the epochs of transitions of the modulating Markov chain, which is an MRW. We use a Wiener-Hopf factorization for this MRW, which also gives results for the busy-period of the storage process, aside from direct results for the unsatisfied demand. Our analysis allows for the Markov chain to have infinite state space and we derive the time dependent as well as the steady state behavior of both the storage level and the unsatisfied demand.

To a great extent, the book's contents that have been summarily described constitute an updated review with many new insights of research work started twenty years ago at Cornell University and originally published in [81, 82, 88–90, 93–95, 112]. António Pacheco and L.C. Tang are grateful respectively to Instituto Superior Técnico and the National University of Singapore for their support to study at the School of Operations Research and Industrial Engineering of Cornell University with Professor N.U. Prabhu as their PhD thesis advisor, and are very honored to have been his last two PhD students. We would like to thank our friends and colleagues from whom we have benefited greatly through some of their earlier works and discussions. This includes, in particular, Dr. Yixin Zhu and Arnold Buss, who have worked on similar problems in the 1980's.

António Pacheco
CEMAT and Instituto Superior Técnico
Technical University of Lisbon

Loon Ching Tang
Department of Industrial and Systems Engineering
National University of Singapore

Narahari U. Prabhu
School of Operations Research and Information Engineering
Cornell University

Contents

Preface v

1. Recurrence and Regeneration 1

2. Markov Renewal and Markov-Additive Processes 7

2.1 Introduction 7
2.2 Markov Renewal Processes: Basic Definitions 10
2.3 Elementary Properties 11
2.4 The Number of Counts 14
2.5 The Markov Renewal Equation 19
2.6 Limit Theorems 21
2.7 The Semi-Markov Process 22
2.8 Markov-Additive Processes: Basic Definitions 23
2.9 A Matrix Equation 26
2.10 The Markov-Poisson Process 28
2.11 Markov-Compound Poisson Processes; A Special Case . . 32
2.12 Markov-Compound Poisson Processes; General Case . . . 34
2.13 The Use of Infinitesimal Generators 36

3. Theory of Semiregenerative Phenomena 39

3.1 Introduction 39
3.2 Basic Definitions 41
3.3 Semirecurrent Phenomena 43
3.4 A Continuous-Time Semiregenerative Phenomenon 50
3.5 Semiregenerative Phenomena With Finite Label Set . . . 56

4. Theory of Markov Random Walks 61

4.1 Introduction 61
4.2 Definitions and Basic Properties 64
4.3 Degenerate Markov Random Walks 67
4.4 Time-Reversed MRW, Extrema, and Semirecurrent Sets . 72
4.5 Fluctuation Theory for MRW 78
4.6 The Case Where Means Exist 81
4.7 Renewal Equation 85
4.8 Wiener-Hopf Factorization 87
4.9 First Exit Time From a Bounded Interval 91

5. Limit Theorems for Markov Random Walks 93

5.1 Introduction 93
5.2 A Sequence of Normalized Increments 94
5.3 Limit Theorems 97
5.4 Application 100

6. Markov-Additive Processes of Arrivals 103

6.1 Introduction 104
6.2 Univariate MAPs of Arrivals 105
6.3 MAPs of Arrivals 109
6.4 Some Properties of MAPs of Arrivals 114
6.5 Some Properties of MAPs 124
6.6 Transformations of MAPs of Arrivals 128
6.7 Markov-Bernoulli Recording of MAPs of Arrivals 134

7. Markov-Modulated Single Server Queueing Systems 141

7.1 Introduction 141
7.2 Preliminary Results 143
7.3 Waiting and Idle Times 146
7.4 Markov-modulated M/M/1 Queue 151

8. A Storage Model for Data Communication Systems 155

8.1 Introduction 155
8.2 The Model 159
8.3 The Actual and the Unsatisfied Demands 163
8.4 The Inverse of the Demand 167
8.5 The Busy-Period Process 170

8.6 Some More Notations and Results 175
8.7 Laplace Transform of the Busy Period 178
8.8 The Unsatisfied Demand and Demand Rejection Rate . . 180
8.9 The Storage Level and Unsatisfied Demand 182
8.10 The Steady State 184

9. A Markovian Storage Model 189

9.1 Introduction 189
9.2 Preliminary Results 193
9.3 The Imbedded MRW 195
9.4 The Main Results 201

Bibliography 207

Index 215

Chapter 1

Recurrence and Regeneration

In its most basic form the concept of recurrence was first formulated by H. Poincaré in his theory of dynamical systems, specifically in the following statement quoted by Chandrasekhar [24]:

> In a system of material particles under the influence of forces which depend only on the spatial coordinates, a given initial state must, in general, recur, not exactly, but to any desired degree of accuracy, infinitely often, provided the system always remains in the finite part of the phase space.

This statement caused controversies between classical and statistical physicists, but was clarified by M.V. Smoluchowski and P. and T. Ehrenfest in probabilistic terms; for historical remarks on this subject see Kac [47, 48].

An equivalent notion is that of regeneration, the first clear formulation of which was given by Palm [83]. Roughly speaking, the idea is that a random process sometimes possesses a set of points (called regeneration points) such that at each such point the process starts from scratch, that is, it evolves anew independently of its past. For a Markov process every point is a regeneration point.

A closely related concept is that of recurrent events, for which W. Feller developed a systematic theory in 1949-1950. A recurrent event is an attribute of a sequence of repeated trials, indicating a pattern such that after each occurrence of this pattern the trials start from scratch. An elementary example is equalization of the cumulated number of heads and tails in coin tossing and success runs in Bernoulli trials, but perhaps the most important application of the theory is to time-homogeneous Markov chains with a countable state space (Example 1.1 below), where the successive returns

[1]This chapter is an updated version of material included in N.U. Prabhu's Technical Report: 'Random Walks, Renewal Processes and Regenerative Phenomena' (June 1994), Uppsala University, Department of Mathematics.

of the chain to a fixed state constitute a recurrent event (this idea had in fact been used earlier by W. Doeblin in 1938). The preceding definition of recurrent events is unsatisfactory because it does not distinguish the pattern (or rather, the phenomenon) from the events that constitute it, and moreover, it does not indicate a possible extension of the concept to continuous time. K.L. Chung's 1967 formulation of repetitive patterns suffers from the same drawbacks. As a consequence, in applied literature one very often finds loose descriptions of recurrent phenomena in discrete as well as continuous time. A continuous time theory was proposed by M.S. Bartlett in 1953, but it deals with a particular case which could be treated by other techniques.

Replacing Feller's framework of an infinite sequence of trials with a countable set of outcomes by a sequence $\{X_n, n \geq 1\}$ of random variables taking values in $S = \{x_1, x_2, \ldots\}$ we may describe a recurrent phenomenon as follows. Let $\sigma\{X_1, X_2, \ldots, X_n\}$ be the σ-field induced by $X_1, X_2, \ldots, X_n \, (n \geq 1)$. Denote by Z_n a random variable taking values in $\{0, 1\}$ and such $Z_0 = 1$ a.s. The sequence $\{Z_n, n \geq 0\}$ is a recurrent phenomenon if it satisfies the following conditions:

(a) For each $n > 1$,

$$\{Z_n = 1\} \in \sigma\{X_1, X_2, \ldots, X_n\}. \tag{1.1}$$

(b) For $\alpha \in \{0, 1\}$ and $m \geq 1, n \geq 1$,

$$\begin{aligned} &P\{X_1 = x_1, X_2 = x_2, \ldots, X_{m+n} = x_{m+n}, Z_{m+n} = \alpha | Z_m = 1\} \\ &\quad = P\{X_1 = x_1, X_2 = x_2, \ldots, X_m = x_m | Z_m = 1\} \cdot \\ &\qquad P\{X_1 = x_{m+1}, X_2 = x_{m+2}, \ldots, X_n = x_{m+n}, Z_n = \alpha\}. \end{aligned} \tag{1.2}$$

However, this definition is really that of a recurrent process $\{X_n\}$, namely, a process which has imbedded in it a recurrent phenomenon.

While it is true that recurrent phenomena rarely occur by themselves, it is possible to define a recurrent phenomenon $\{Z_n, n \geq 0\}$ rather simply as follows. Let $\{Z_n, n \geq 0\}$ be a sequence of random variables on a probability space $(\Omega, \mathcal{F}, P)$, taking values in $\{0, 1\}$ and such that $Z_0 = 1$ a.s. and

$$\begin{aligned} &P\{Z_{n_1} = Z_{n_2} = \ldots = Z_{n_r} = 1\} \\ &\qquad = P\{Z_{n_1} = 1\} P\{Z_{n_2 - n_1} = 1\} \cdots P\{Z_{n_r - n_{r-1}} = 1\} \end{aligned} \tag{1.3}$$

for every set of integers n_i such that

$$0 < n_1 < n_2 < \cdots < n_r \quad (r \geq 1).$$

Example 1.1. Suppose $\{X_n, n \geq 0\}$ is a time-homogeneous Markov chain on the state space $S = \{0, 1, 2, \ldots\}$. For $a \in S$, assume $X_0 = a$ a.s., and define the random variables Z_n as follows:

$$Z_n = \begin{cases} 1 & \text{if } X_n = a \\ 0 & \text{otherwise.} \end{cases}$$

It is seen that the property (1.3) is satisfied and hence $\{Z_n, n \geq 0\}$ is a recurrent phenomenon, with

$$u_n = P\{Z_n = 1\} = P_{aa}^{(n)},$$

where $P_{ij}^{(n)}$ $(i, j \in S, n \geq 0)$ are the n-step transition probabilities of the chain. Conversely, the sequence $\{u_n\}$ associated with a recurrent phenomenon can arise only in this way (K.L. Chung). Actually we can do better, as indicated in the following.

Example 1.2. Let $\{X_k, k \geq 1\}$ be a sequence of independent and identically distributed random variables on the state space $\{1, 2, \ldots\}$. Denote $S_0 \equiv 0, S_k = X_1 + X_2 + \ldots + X_k \ (k \geq 1)$ and

$$R = \{n \geq 0 : S_k = n \text{ for some } k \geq 0\}.$$

That is, R is the range of the renewal process $\{S_k, k \geq 0\}$. For $n \geq 0$ let

$$Z_n = \begin{cases} 1 & \text{if } n \in R \\ 0 & \text{otherwise.} \end{cases}$$

Then it can be shown that $\{Z_n, n \geq 0\}$ is a recurrent phenomenon. Conversely, it turns out that every recurrent phenomenon is the indicator function of the range of a renewal process induced by a distribution on the positive integers. Then X_k are identified as the recurrence times (Poincaré cycles).

The definition (1.3) suggests the appropriate extension of the concept of recurrent phenomena to continuous time, but we might now prefer the term regenerative to recurrent as a matter of fashion. Thus the family of random variables $\{Z(t), t \geq 0\}$ is a regenerative phenomenon if for each t, $Z(t)$ takes values in $\{0, 1\}$ and

$$\begin{aligned} P\{Z(t_1) = Z(t_2) = \ldots = Z(t_r) = 1\} \\ = P\{Z(t_1) = 1\}P\{Z(t_2 - t_1) = 1\} \ldots P\{Z(t_r - t_{r-1}) = 1\} \quad (1.4) \end{aligned}$$

whenever $0 < t_1 < t_2 < \ldots < t_r \ (r \geq 1)$. The definitions (1.3) and (1.4) were proposed by J.F.C. Kingman in 1963 and the theory of continuous-time regenerative phenomena was developed by him in a series of papers published during 1964-1970. He also gave a streamlined account of the theory [55].

Example 1.3. Let $\{B(t), t \geq 0\}$ be a Brownian motion on $(-\infty, \infty)$ and define

$$Z(t) = \begin{cases} 1 & \text{if } B(t) = 0 \\ 0 & \text{if } B(t) \neq 0 \end{cases}$$

Thus $Z(t)$ is the indicator function of the state 0. Clearly, the property (1.4) is satisfied and $\{Z(t), t \geq 0\}$ is a regenerative phenomenon. However, we have

$$p(t) = P\{Z(t) = 1\} = 0.$$

Kingman considers regenerative phenomena satisfying the condition

$$p(t) \to 1 \text{ as } t \to 0^+. \tag{1.5}$$

If this condition is satisfied, then $p(t) > 0$ for all $t \geq 0$, and we may take $p(0) = 1$. Such p-functions and the corresponding phenomena are called standard.

Example 1.4. Let $\{X(t), t \geq 0\}$ be a time-homogeneous Markov process on the state space $S = \{0, 1, 2, \ldots\}$. Let $a \in S$, $X_0 = a$ a.s. and for $t \geq 0$ define

$$Z(t) = \begin{cases} 1 & \text{if } X(t) = a \\ 0 & \text{otherwise.} \end{cases}$$

Then, as in Example 1.1, it is seen that $\{Z(t), t \geq 0\}$ is a regenerative phenomenon, with

$$p(t) = P\{Z(t) = 1\} = P_{aa}(t),$$

where $P_{ij}(t)\,(i, j \in S, t \geq 0)$ is the transition probability function of the process. The usual assumptions on these functions lead to the fact that the phenomenon is standard. Unlike the situation in discrete time, the function p of a standard regenerative phenomenon need not arise only in this manner. Kingman [55] investigates the so-called Markov characterization problem, namely, the problems of establishing a criterion for a standard p-function to be a diagonal element of some Markov transition matrix.

Example 1.2 shows the connection between recurrent phenomena and renewal processes. In continuous time a similar connection exists between regenerative phenomena and Lévy processes with non-decreasing sample functions (subordinators); the survey paper by Fristedt [39] sheds more light on this connection. Maisonneuve [67] has studied regenerative phenomena from a slightly different point of view, and a theory of stochastic regenerative processes developed earlier by Smith [107, 108] has considerably influenced research workers in this area.

For the theory of recurrent phenomena, despite the drawbacks in its formulation indicated above, the best sources are still Feller's (1949) paper [34] and his book ([35], Chapter XIII).

We shall use the term regenerative phenomena to cover the discrete time as well as the continuous-time cases. For a more recent treatment see Prabhu ([92], Chapter 9).

Semiregenerative phenomena are regenerative phenomena that have, in addition, a Markov component. Such phenomena are important in the study of Markov-additive processes, and arise in several important applications. In the following chapters we study semiregenerative phenomena, including, in particular, their connections with Markov-additive processes.

Chapter 2

Markov Renewal and Markov-Additive Processes

The earliest Markov-modulated processes studied in the literature are Markov-additive processes. In this chapter we investigate the properties of some of these processes which are of importance in applied probability. We also review the basic concepts and main properties of Markov renewal processes, which are discrete time Markov-additive processes that may be viewed as a family of Markov-dependent renewal processes.

2.1 Introduction

Loosely speaking, a Markov-modulated process is a bivariate Markov process (X, J) such that J is also a Markov process and the future behavior of X or some function of it (like its increments) is independent of the past behavior of the process given the current state of J. These properties give a special role to the component J, the so called *Markov component*, whose properties may determine to a great extent the properties of the process (X, J) and which makes it possible to strengthen the randomness of the process X in a very natural way.

In many applications J represents an extraneous factor, such as the environment, which may be non-observable. This is the reason why some authors call Hidden Markov Models (HMMs) to Markov-modulated processes. The variety of applications of Markov-modulated processes is well illustrated by the following selected short list of subjects of recent applica-

[2]This chapter is an updated version of material included in the article by N.U. Prabhu, 'Markov-renewal and Markov-additive processes – a review and some new results', in *Analysis and Geometry 1991* (Korea Adv. Inst. Sci. Tech., Taejŏn), pp. 57–94, 1991, and contents are reproduced with permission of the publisher, the Korea Advanced Institute of Science and Technology.

tions: branching processes [4], DNA analysis [79], finance [32], music [113], plasma turbulence [20], polymerase chain reaction [59], risk theory [106], and word occurrence [84].

The earliest Markov-modulated processes studied in the literature are Markov-additive processes. A Markov-additive processes (X, J) is a two-dimensional Markov process such that the component X is additive, the distribution of its increments being modulated by (or depending on) J. The pioneering study of Markov-additive processes (MAPs) is due to Çinlar [28, 29], who developed the theory along general lines, and a recent treatment of Markov-additive processes on general state spaces is given by Breuer [21, 22]. MAPs are generalizations of Lévy processes, which have stationary independent increments, since when the state space of J consists of only one element the additive component X of an MAP (X, J) has independent increments.

MAPs occur in many probability models; in particular, in Markov-modulated queueing systems the basic processes are Markov-Poisson and Markov-compound Poisson processes as it is highlighted in chapters 7–9 of the book. The Markov-modulated Brownian motion is another MAP that has become increasingly popular in describing various risk and queueing models, as illustrated, e.g., in [11, 12, 32, 49, 117]. Markov subordinators (Markov-additive processes whose additive component takes nonnegative values) play a key role in the theory of semiregenerative phenomena, as developed by Prabhu [88, 90] and reviewed in the next chapter.

Starting from Sec. 2.8, we investigate the properties of special cases of MAPs which are of importance in applied probability, including the so called MMPP (Markov-modulated Poisson process), in which the rate of occurrence of Poisson events changes instantaneously with the changes of state of the modulating Markov chain, along with other Markov-compound Poisson processes. We also illustrate the use of infinitesimal generators for MAPs. Some additional properties of MAPs will be given in Chap. 6, where the important class of MAPs of arrivals [81] will be studied. These are MAPs whose additive components take values in the nonnegative (may be multidimensional) integers, so that their increments may be interpreted as corresponding to arrivals, the standard example being that of different classes of arrivals into a queueing system.

In the discrete time case MAPs are Markov random walks (MRWs), whose theory is reviewed in chapters 4 and 5. MRWs specialize into Markov renewal processes (MRPs) when the increments of the additive component are nonnegative. Although there is now a vast literature on Markov renewal theory and its applications, the pioneering paper by Çinlar [27] is an extensive and authoritative survey, containing historical references on the subject, that still remains the single major reference on the subject. Streamlined treatments of MRPs will be fond in Çinlar ([30], Chapter 10) and Prabhu [89]. More recent books on applied probability contain brief sketches of the theory with a view to applications to probability models, including in particular [12, 31, 45, 46, 57, 60].

In the first part of this chapter, namely in sections 2.2–2.7, we present a review of the literature on Markov renewal theory. This is not meant to be a historical survey, nor does it review applications. Rather, this is an exposition of the basic concepts and main properties of Markov renewal processes. Our approach is somewhat different from the ones used in standard treatments of the subject. We view a Markov renewal process (MRP) as essentially an extension of the standard renewal process $\{S_n, n \geq 0\}$ in the sense that it is modulated by an underlying Markov chain $J = \{J_n, n \geq 0\}$ on a countable state space E.

Thus an MRP is a two-dimensional Markov process $\{(S_n, J_n), n \geq 0\}$ where the increments in the first component have an arbitrary distribution (rather than an exponential density as in a Markov process). In such a treatment, the terminology (such as transition kernel and sojourn times) and the classification of states are motivated by this so-called semi-Markov process. This process might have provided the historic motivation for the study of this subject area, but the treatment has caused a certain amount of confusion in the literature, especially in regard to the terms Markov renewal processes and semi-Markov processes.

Our approach makes it clear that an MRP is a family of Markov-dependent renewal processes indexed by E. The corresponding family $\{N_k(t), k \in E\}$ of renewal counting processes gives rise in a natural fashion to the so-called Markov renewal equation. We establish the connection between the existence and uniqueness of the solution of this equation and the finiteness of the total number of renewals, namely, $N(t) = \sum_{k\in E} N_k(t)$. Standard treatments of the Markov renewal equation are purely analytical.

Feller ([35], Chapter XIII) has given a lucid treatment of recurrent events and renewal theory. It is natural that connected with Markov renewal theory there should be a theory of semirecurrent events (or to use a preferred term, a semirecurrent phenomenon). The connection between Markov renewal processes and semirecurrent phenomena, developed by Prabhu [88], will be reviewed in the next chapter.

2.2 Markov Renewal Processes: Basic Definitions

We are given a probability space $(\Omega, \mathcal{F}, P)$ and denote $\mathbf{R}_+ = [0, \infty)$, $E =$ a countable set, and $\mathbf{N}_+ = \{0, 1, 2, \ldots\}$.

Definition 2.1. A Markov renewal process $(S, J) = \{(S_n, J_n), n \geq 0\}$ is a Markov process on the state space $\mathbf{R} \times E$ whose transition distribution measure has the property

$$P\{(S_{m+n}, J_{m+n}) \in A \times \{k\} | (S_m, J_m) = (x, j)\}$$
$$= P\{(S_{m+n} - S_m, J_{m+n}) \in (A - x) \times \{k\} | J_m = j\} \qquad (2.1)$$

for every $j, k \in E$, $x \in \mathbf{R}_+$ and a Borel subset A of $\mathbf{R}_+$.

We shall only consider the homogeneous process for which the second probability in (2.1) does not depend on m. In this case, denoting this probability as $Q_{jk}^{(n)}\{A - x\}$ we see that

$$Q_{jk}^{(n)}\{A\} = P\{(S_n, J_n) \in A \times \{k\} | J_0 = j\}, \qquad (2.2)$$

where we shall write $Q_{jk}\{A\}$ for $Q_{jk}^{(1)}\{A\}$. For each $j \in E$, $n \geq 0$, $Q_{jk}^{(n)}$ is a distribution measure. We have

$$Q_{jk}^{(n)}\{A\} \geq 0, \qquad \sum_{k \in E} Q_{jk}^{(n)}\{\mathbf{R}_+\} \leq 1 \qquad (2.3)$$

$$Q_{jk}^{(0)}\{A\} = \begin{cases} 1 & \text{if } j = k \text{ and } 0 \in A \\ 0 & \text{otherwise.} \end{cases} \qquad (2.4)$$

For the Chapman-Kolmogorov equations we have

$$Q_{jk}^{(m+n)}\{A\} = \sum_{l \in E} \int_{0-}^{\infty} Q_{jl}^{(m)}\{dx\} Q_{lk}^{(n)}\{A - x\} \qquad (m, n \geq 0). \qquad (2.5)$$

The initial measure is given by

$$P\{(S_0, J_0) \in A \times \{j\}\} = \begin{cases} a_j & \text{if } 0 \in A \\ 0 & \text{otherwise} \end{cases} \qquad (2.6)$$

and the finite dimensional distributions of the process are given by

$$P\{(S_{n_t}, J_{n_t}) \in A_t \times \{k_t\} (1 \leq t \leq r)\}$$
$$= \sum_{j \in E} a_j \int_{x_t \in A_t (1 \leq t \leq r)} Q_{jk_1}^{(n_1)}\{dx_1\} Q_{k_1 k_2}^{(n_2 - n_1)}\{dx_2 - x_1\}$$
$$\cdots Q_{k_{r-1} k_r}^{(n_r - n_{r-1})}\{dx_r - x_{r-1}\} \qquad (2.7)$$

for $0 \leq n_1 < n_2 < \cdots < n_r$ and $k_1, k_2, \ldots, k_r \in E$. We shall use the abbreviation MRP for a Markov renewal process. Also, we shall denote by P_j and E_j the conditional probability and the expectation given $J_0 = j$.

2.3 Elementary Properties

For the marginal process $J = \{J_n, n \geq 0\}$ we have the following.

Proposition 2.1. *The marginal process J of the MRP is a Markov chain on the state space E with the transition probabilities*

$$P_{jk}^{(n)} = Q_{jk}^{(n)}\{\mathbf{R}_+\} \quad (n \geq 0). \tag{2.8}$$

Proof. Since (S, J) is a Markov process and has the property (2.1), we find that

$$\begin{aligned} P\{J_{m+n} = k | (S_m, J_m) = (x, j)\} &= P\{J_n = k | J_0 = j\} \\ &= Q_{jk}^{(n)}\{\mathbf{R}_+\}. \end{aligned}$$

This shows that $\{J_n, n \geq 0\}$ is a Markov chain with the transition probabilities given by (2.8). □

Let $X_n = S_n - S_{n-1}(n \geq 1)$. The X_n are called the *lifetimes* of the process. Clearly, $X_n \geq 0$ a.s. $(n \geq 1)$. We have the following.

Proposition 2.2. *For $r \geq 1$ we have*

$$\begin{aligned} &P\{(X_t, J_t) \in A_t \times \{k_t\}(1 \leq t \leq r)\} \\ &\qquad = \sum_{j \in A} a_j Q_{jk_1}\{A_1\} Q_{k_1 k_2}\{A_2\} \cdots Q_{k_{r-1} k_r}\{A_r\} \end{aligned} \tag{2.9}$$

for $0 \leq n_1 < n_2 < \cdots < n_r$ and $k_1, k_2, \ldots, k_r \in E$.

Proof. The described result follows from (2.7). □

In terms of conditional probabilities, the property (2.9) expresses the conditional independence of the lifetimes $X_1, X_2, \ldots, X_r$, given $J_1, J_2, \ldots, J_r$. If $E = \{j\}$, (2.9) shows that these lifetimes are mutually independent and consequently the MRP reduces to a renewal process induced by the distribution Q_{jj}. More generally we have the following.

Proposition 2.3. *Let C be a fixed subset of E, $N_0 = 0$ a.s., and for $r \geq 1$,*

$$N_r = \inf\{n > N_{r-1} : J_n \in C\} \tag{2.10}$$

Then $\{(N_r, S_{N_r}, J_{N_r}\}, r \geq 0\}$ is an MRP on the state space $\mathbf{N}_+ \times \mathbf{R}_+ \times C$, the renewal component being the two-dimensional process (N_r, S_{N_r}).

Proof. The random variables $N_r(n \geq 1)$ are stopping times for (S, J). Using the strong Markov property of the process we find that

$$\begin{aligned} P\{(N_{r+1}, &S_{N_{r+1}}, J_{N_{r+1}}) \in \{n\} \times A \times \{k\}| \\ &\qquad (N_t, S_{N_t}, J_{N_t}) = (m_t, x_t, j_t), (1 \leq t \leq r)\} \\ &= P\{(N_{r+1} - N_r, S_{N_{r+1}} - S_{N_r}, J_{N_{r+1}}) \\ &\qquad \in \{n - m_r\} \times (A - x_r) \times \{k\} | J_{N_r} = j_r\}, \end{aligned}$$

which establishes the desired property. □

In the above result, if $C = \{k\}$, we conclude that $\{(N_r, S_{N_r}), r \geq 0\}$ is a renewal process on $\mathbf{N}_+ \times \mathbf{R}_+$. Of special interest is the marginal process $\{S_{N_r}\}$, which is also a renewal process. Denote $S_r^k = S_{N_r} (r \geq 0)$. We shall call $S^k = \{S_r^k, r \geq 0\}$ its *kth imbedded renewal process.* The MRP (S, J) may be viewed as a family of (Markov-dependent) renewal processes.

Let us denote by

$$\mathcal{R} = \{(t, j) \in \mathbf{R}_+ \times E : (S_n, J_n) = (t, j) \text{ for some } n \geq 0\} \tag{2.11}$$

the *observed range* of the (S, J) process,

$$N_k(t) = \#\{n > 0 : (S_n, J_n) \in \mathcal{R} \cap ([0, t] \times \{k\})\} \tag{2.12}$$

and

$$N(t) = \sum_{k \in E} N_k(t) \leq \infty. \tag{2.13}$$

Here $N_k(t)$ is the number of visits to k during the interval $0 \leq S_n \leq t$, and $N(t)$ the total number of visits to all $k \in E$ during this interval. Concerning the random variable $N_k(t)$ we have the following.

Proposition 2.4. *Suppose that*

$$P_j\{S_0 = S_1 = S_2 = \cdots\} = 0 \textit{ for all } j \in E. \tag{2.14}$$

Then $N_k(t) < \infty$ a.s. and

$$U_{jk}(t) = E_j[N_k(t)] = \sum_{n=1}^{\infty} Q_{jk}^{(n)}\{[0, t]\} < \infty \tag{2.15}$$

for all $t \in [0, \infty)$ and $k \in E$. Also, $U_{jk}(t)$ satisfies the integral equation

$$U_{jk}(t) = Q_{jk}\{[0, t]\} + \sum_{l \in E} \int_{0-}^{t} Q_{jl}\{ds\} U_{lk}(t - s). \tag{2.16}$$

Proof. The assumption (2.14) implies that on $\{J_0 = j\}$ the distribution inducing the renewal process S^k is not concentrated at 0. The first two results follow from standard renewal theory. To prove (2.16) we note that

$$U_{jk}(t) = E_j\left[\sum_{n=1}^{\infty} \mathbf{1}_{\{S_n \le t, J_n = k\}}\right] = \sum_{n=1}^{\infty} Q_{jk}^{(n)}\{[0,t]\}.$$

Therefore

$$\begin{aligned} U_{jk}(t) &= Q_{jk}\{[0,t]\} + \sum_{n=1}^{\infty}\sum_{l\in E}\int_{0-}^{t} Q_{jl}\{ds\}Q_{lk}^{(n)}\{[0,t-s]\} \\ &= Q_{jk}\{[0,t]\} + \sum_{l\in E}\int_{0-}^{t} Q_{jl}\{ds\}U_{lk}(t-s) \end{aligned}$$

as was required to be proved. □

We note that for each $j \in E$, $U_{jk}(t)$ is bounded over finite intervals, while for each $t \in [0,\infty)$, $U_{jk}(t)$ is bounded as a function of j, since

$$U_{jk}(t) \le 1 + U_{kk}(t). \tag{2.17}$$

The question of uniqueness of U as a solution of the integral equation (2.16) will be discussed in Sec. 2.5.

In order to derive limit theorems for the process S^k we need to obtain its mean lifetime $\mu_{jk} = E_j(S_1^k)$. We note that $S_1^k = S_{N_1}$, where

$$N_1 = \inf\{n > 0 : J_n = k\}. \tag{2.18}$$

That is, N_1 is the hitting time of the state k for the marginal chain J.

Lemma 2.1. *If μ_{jk} is finite, then*

$$\mu_{jk} = \sum_{l\in E} {}^kP_{jl}^{*} E_l(X_1), \tag{2.19}$$

where

$${}^kP_{jl}^{*} = \sum_{n=0}^{\infty} P_j\{N_1 > n, J_n = l\}. \tag{2.20}$$

Proof. We have

$$\begin{aligned}
\mu_{jk} &= E_j(S_{N_1}) \\
&= \sum_{n=1}^{\infty} E_j(X_1 + X_2 + \cdots + X_{N_1}; N_1 = n) \\
&= \sum_{n=1}^{\infty} \sum_{m=1}^{n} E_j(X_m; N_1 = n) \\
&= \sum_{m=1}^{\infty} \sum_{n=m}^{\infty} E_j(X_m; N_1 = n) \\
&= \sum_{m=1}^{\infty} E_j(X_m; N_1 \geq m) \\
&= \sum_{m=1}^{\infty} \sum_{l \in E} E_j(X_m; N_1 > m-1, J_{m-1} = l) \\
&= \sum_{l \in E} \sum_{m=1}^{\infty} P_j\{N_1 > m-1, J_{m-1} = l\} E_j(X_m | J_{m-1} = l) \quad (2.21)
\end{aligned}$$

since the event $\{N_1 > m-1\}$ depends only on $J_1, J_2, \ldots, J_{m-1}$. The desired result follows from (2.21) since by time-homogeneity

$$E_j(X_m | J_{m-1} = l) = E_l(X_1).$$

□

2.4 The Number of Counts

We now consider the random variable $N(t)$ defined by (2.13). Let

$$q_j^{(n)}(t) = \sum_{k \in E} Q_{jk}^{(n)}\{[0,t]\} = P_j\{N(t) \geq n\}. \quad (2.22)$$

For fixed $j \in E$, $t \in \mathbf{R}_+$, the sequence $\{q_j^{(n)}(t), n \geq 0\}$ is monotone nonincreasing, and for its limit we have

$$q_j(t) = \lim_{n \to \infty} q_j^{(n)}(t) = P_j\{N(t) = \infty\}. \quad (2.23)$$

Theorem 2.1. *$q = \{q_j(t), j \in E\}$ is the maximal solution of the integral equation*

$$x_j(t) = \sum_{l \in E} \int_{0-}^{t} Q_{jl}\{ds\} x_l(t-s), \qquad j \in E, t \in \mathbf{R}_+ \quad (2.24)$$

with $0 \leq x_j \leq 1$, $j \in E$, in the sense that $x_j(t) \leq q_j(t)$.

Proof. From (2.4) and (2.5) we find that for all $t \in \mathbf{R}_+$,

$$q_j^{(0)}(t) = 1 \tag{2.25}$$

and

$$q_j^{(n+1)}(t) = \int_{l \in E} \int_{0-}^{t} Q_{jl}\{ds\} q_l^{(n)}(t-s) \quad (n \geq 0). \tag{2.26}$$

By the dominated convergence theorem we find from (2.26) that q satisfies the integral equation (2.24). If $x = \{x_j(t), j \in E\}$ is a solution of (2.24) with $0 \leq x_j \leq 1$, $j \in E$, then $x_j(t) \leq q_j^{(0)}(t)$ and

$$x_j(t) \leq \sum_{l \in E} \int_{0-}^{t} Q_{jl}\{ds\} = \sum_{l \in E} Q_{jl}\{[0,t]\} = q_j^{(1)}(t).$$

By induction it follows that

$$x_j(t) \leq q_j^{(n)}(t) \quad (n \geq 0).$$

In the limit as $n \to \infty$, this gives $x_j(t) \leq q_j(t)$ for $j \in E$, $t \in \mathbf{R}_+$, as was required to be proved. □

Theorem 2.2. *$N(t) < \infty$ for all $t \in \mathbf{R}_+$ a.s. iff the only solution of the integral equation (2.24) with $0 \leq x_j \leq 1$, $j \in E$, is given by $x_j(t) = 0$, $j \in E$, $t \in \mathbf{R}_+$.*

Proof. If $x = 0$ is the only solution of (2.24), then since q is a solution, we must have $q_j(t) = P_j\{N(t) = \infty\} = 0$ for $j \in E$, $t \in \mathbf{R}_+$. Conversely, if $q = 0$, then since $x \leq q$, we must have $x = 0$. □

We shall also consider the random variable

$$N = \#\{n > 0 : (S_n, J_n) \in \mathcal{R}\} \tag{2.27}$$

where $\mathcal{R}$ is the observed range defined by (2.11). Here N is the maximum number of observed transitions, and the process terminates at time N. Clearly, $N = N(\infty)$. We can then write

$$Q_{jk}^{(n)}\{A\} = P\{(S_n, J_n) \in A \times \{k\}, n \leq N | J_0 = j\} \tag{2.28}$$

and

$$\sum_{k \in E} Q_{jk}^{(n)}\{\mathbf{R}\} = P_j\{N \geq n\}. \tag{2.29}$$

This explains the second set of inequalities in (2.3). Let us also write

$$\sigma = \sup_{0 \leq n \leq N} S_n. \tag{2.30}$$

We note that if $N < \infty$, then $\sigma = S_N < \infty$. However, σ can be finite even when $N = \infty$. Accordingly we need to consider the following three cases:

(a) $\sigma = \infty$
(b) $\sigma < \infty, N < \infty$
(c) $\sigma < \infty, N = \infty$.

We have the following.

Theorem 2.3. *In cases (a) and (b), $N(t) < \infty$ for all $t \in \mathbf{R}_+$, while in case (c) we have $N(t) = \infty$ for some $t \in \mathbf{R}_+$.*

Proof. Clearly, $\{\sigma \leq t\} = \{N(t) = N\}$. Therefore if $\sigma = \infty$, then $N(t) < N$ for all $t \in \mathbf{R}_+$, where $N = \infty$. If $\sigma < \infty$, $N < \infty$, then $N(t) = N$ for $t \geq \sigma$, so that $N(t) < \infty$ for all $t \in \mathbf{R}_+$. Finally, if $\sigma < \infty$, $N = \infty$, we find that $N(t) = \infty$ for $t \geq \sigma$. □

Remark 2.1. In the standard renewal theory only the cases (a) and (b) arise, corresponding respectively to nonterminating and terminating renewal processes. The remaining case (c) is an additional feature of Markov renewal processes.

Remark 2.2. Instead of (2.30) if we define

$$L = \sup_{n \geq 0} S_n \tag{2.31}$$

then we see that $\sigma = L \leq \infty$ if $N = \infty$, while $\sigma < L = \infty$ if $N < \infty$. More specifically

$$\begin{aligned} \sigma = L = \infty \quad & \text{in case (a)} \\ \sigma < L = \infty \quad & \text{in case (b)} \\ \sigma = L < \infty \quad & \text{in case (c)} \end{aligned}$$

Theorem 2.3 implies that $N(t) < \infty$ for all $t \in \mathbf{R}_+$ iff $L = \infty$. However, this statement does not distinguish between cases (a) and (b), namely between nonterminating and terminating Markov renewal processes.

Theorem 2.4. *(a) If the marginal chain J is persistent, then $N(t) < \infty$ for all $t \in \mathbf{R}_+$ a.s.*
(b) In the chain J if the probability of forever remaining in the transient states is zero, then $N(t) < \infty$ for all $t \in \mathbf{R}_+$ a.s.

Proof. *(a).* If the J chain is persistent, then for the kth renewal process we have $\sigma^k = \sup_{0 \leq r \leq N^k} S_r^k = \infty$, where

$$N^k = \#\{n > 0 : J_n = k, (S_n, J_n) \in \mathcal{R}\}.$$

Since $\sigma \geq \sigma^k = \infty$ it follows from Theorem 2.3 that $N(t) < \infty$ for all $t \in \mathbf{R}_+$ a.s. on $\{J_0 = k\}$. Since this is true for all $k \in E$, we obtain the desired result.

(b). From (2.22) and (2.29) we find that

$$q_j^{(n)}(\infty) = \sum_{k \in E} P_{jk}^{(n)} = P_j\{N \geq n\} \tag{2.32}$$

where the $P_{jk}^{(n)}$ are the transition probabilities of the J chain, as defined by (2.8). Also,

$$q_j(\infty) = \lim_{n \to \infty} q_j^{(n)}(\infty) = P_j\{N = \infty\}, \tag{2.33}$$

so that $q_j(\infty)$ is the probability that J remains forever in E. This argument applies equally well to the subset of E formed by the transient states and if $q_j(\infty) = 0$ for all transient states, then $q_j(t) \leq q_j(\infty) = 0$. Using Theorem 2.2 and (a), we conclude that $N(t) < \infty$ for all $t \in \mathbf{R}_+$ a.s. □

Note that from Theorem 2.4 (b), we conclude that if E is finite, then $N(t) < \infty$ for all $t \in \mathbf{R}_+$ a.s.

Example 2.1. We are given $E = \{0, 1, 2\}$ and proper distributions $F_0, F_1, F_{20}, F_{21}, F_{22}$. We define an MRP by its transition distribution measures

$$Q_{00}\{A\} = F_0\{A\}, \quad Q_{01}\{A\} = Q_{02}\{A\} = 0$$

$$Q_{10}\{A\} = 0, \quad Q_{11}\{A\} = \frac{3}{4}F_1\{A\}, \quad Q_{12}\{A\} = 0$$

$$Q_{2k}\{A\} = P_{2k}F_{2k}\{A\} \quad (k = 0, 1, 2),$$

where $P_{20} + P_{21} + P_{22} = 1$ and $P_{22} < 1$. The marginal chain J has the transition probability matrix

$$\begin{bmatrix} 1 & 0 & 0 \\ 0 & 3/4 & 0 \\ P_{20} & P_{21} & P_{22} \end{bmatrix}$$

For this chain the state 0 is absorbing, while the states 1 and 2 are transient.

The sample functions of the MRP are of the following forms:

J_0	Sample path	(N, σ)
0	$(S_0, 0), (S_1, 0), (S_2, 0), \ldots$	$N = \infty, \sigma = \infty$
1	$(S_0, 1), (S_1, 1), \ldots, (S_N, 1)$	$N < \infty, \sigma = S_N < \infty$
2	$(S_0, 2), (S_1, 2), \ldots, (S_m, 2),$ $(S_{m+1}, 0), (S_{m+2}, 0), \ldots$	$N = \infty, \sigma = \infty$
2	$(S_0, 2), (S_1, 2), \ldots, (S_m, 2),$ $(S_{m+1}, 1), (S_{m+2}, 1), \ldots, (S_N, 1)$	$N < \infty, \sigma = S_N < \infty$

We see that from state 0 the process does not terminate a.s. and from state 1 the process terminates a.s. From state 2 the process terminates if it gets into state 1 or else does not terminate if it gets into state 0. Therefore

$$P_0\{\sigma = \infty\} = 1$$

$$P_1\{N = n, \sigma \in A\} = \left(\frac{1}{4}\right)\left(\frac{3}{4}\right)^n F_1^{(n)}\{A\} \ (n \geq 0)$$

so that

$$P_1\{N < \infty, \sigma < \infty\} = 1.$$

Also,

$$P_2\{N = \infty, \sigma = \infty\} = \sum_{n=0}^{\infty} (P_{22})^n P_{20} = \frac{P_{20}}{1 - P_{22}}.$$

Finally, for $n \geq 1$

$$P_2\{N = n, \sigma \in A\} = \sum_{m=0}^{n-1} (P_{22})^m P_{21} \int\int F_{22}^{(m)}\{dx_1\} F_{21}\{dx_2\} \left(\frac{3}{4}\right)^{n-m-1} \cdot F_1^{(n-m-1)}\{A - x_1 - x_2\} \left(\frac{1}{4}\right).$$

This gives

$$\begin{aligned} P_2\{N < \infty, \sigma < \infty\} &= \sum_{n=1}^{\infty} \sum_{m=0}^{n-1} (P_{22})^m P_{21} \left(\frac{3}{4}\right)^{n-m-1} \left(\frac{1}{4}\right) \\ &= \frac{P_{21}}{1 - P_{22}}. \end{aligned}$$

From Theorem 2.3 it follows that

$$P_j\{N(t) < \infty, t \in \mathbf{R}_+\} = 1 \quad (j = 0, 1, 2).$$

Example 2.2. Here $E = \{0, 1, 2, \dots\}$ and

$$Q_{jk}\{A\} = \begin{cases} F_j\{A\} & \text{if } k = j + 1 \\ 0 & \text{otherwise,} \end{cases}$$

where F_j is a proper distribution such that as $n \to \infty$

$$F_j * F_{j+1} * \cdots * F_{j+n-1}\{A\} \to G_j\{A\},$$

G_j being a proper distribution. We have

$$Q_{jk}\{A\} = \begin{cases} F_j * F_{j+1} * \cdots * F_{j+n-1}\{A\} & \text{if } k = j + n \\ 0 & \text{otherwise.} \end{cases}$$

Therefore

$$q_j^{(n)}(t) = \sum_{k \in E} Q_{jk}^{(n)}\{[0,t]\} \to G_j\{[0,t]\}$$

as $n \to \infty$. We conclude that

$$P_j\{N = \infty, \sigma \in A\} = G_j\{A\}$$

and

$$P_j\{N = \infty, \sigma < \infty\} = 1.$$

Here the sample functions are of the form

$$(s_0, j), (s_1, j+1), (s_2, j+2), \ldots .$$

2.5 The Markov Renewal Equation

Let **B** denote the class of nonnegative functions $f_j(t)$, $j \in E$, $t \in \mathbf{R}_+$, such that for every $j \in E$, $f_j(t)$ is bounded over finite intervals, and for every $t \in \mathbf{R}_+$, $f_j(t)$ is bounded. Examples of such functions are (for fixed $k \in E$), the renewal function $U_{jk}(t)$ (as remarked in Sec. 2.3), the transition distribution function $Q_{jk}\{[0,t]\}$ and also

$$\sum_{l \in E} \int_{0-}^{t} Q_{jl}\{ds\} f_l(t-s), \tag{2.34}$$

where $f_l(t)$ belongs to **B**. The Markov renewal equation is the integral equation

$$f_j(t) = g_j(t) + \sum_{l \in E} \int_{0-}^{t} Q_{jl}\{ds\} f_l(t-s) \tag{2.35}$$

where $g_j(t)$ is a known function which belongs to **B**. By Proposition 2.4, the function $U_{jk}(t)$ satisfies (2.35) with $g_j(t) = Q_{jk}\{[0,t]\}$. We shall consider the existence and uniqueness of the solution of (2.35) that belongs to **B**. If $E = \{j\}$, then (2.35) reduces to the renewal equation of standard renewal theory and we know that it has a unique solution. In the general case it would be convenient to introduce the operator notation $Q * f$ to express (2.34). Thus

$$(Q * f)_j(t) = \sum_{l \in E} \int_{0-}^{t} Q_{jl}\{ds\} f_l(t-s) \tag{2.36}$$

It is clear that $Q*f$ also belongs to $\mathbf{B}$, so that we can define the nth iterate $Q^n * f$ by

$$(Q^n * f)_j(t) = (Q * (Q^{n-1} * f))_j(t) \quad (n \geq 1)$$

with $Q^0 * f = f$. We find that

$$(Q^n * f)_j(t) = \sum_{l \in E} \int_{0-}^{t} Q_{jl}^{(n)}\{ds\} f_l(t-s) \quad (n \geq 0).$$

We define the operator U by

$$\begin{aligned}(U * f)_j(t) &= \sum_{n=0}^{\infty} \sum_{l \in E} \int_{0-}^{t} Q_{jl}\{ds\} f_l(t-s) \\ &= \sum_{l \in E} \int_{0-}^{t} U_{jl}\{ds\} f_l(t-s) \end{aligned} \tag{2.37}$$

where we have modified the definition of Sec. 2.3 to

$$U_{jk}\{A\} = \sum_{n=0}^{\infty} Q_{jk}^{(n)}\{A\} < \infty. \tag{2.38}$$

Theorem 2.5. *A particular solution of the Markov renewal equation (2.35) is given by*

$$G_j(t) = \sum_{l \in E} \int_{0-}^{t} U_{jl}\{ds\} g_l(t-s), \tag{2.39}$$

while the general solution is of the form

$$f_j(t) = G_j(t) + h_j(t) \tag{2.40}$$

where h satisfies the equation

$$h_j(t) = \sum_{l \in E} \int_{0-}^{t} Q_{jl}\{ds\} h_l(t-s). \tag{2.41}$$

Proof. The Markov renewal equation (2.35) can be written as

$$f = g + Q * f. \tag{2.42}$$

Substituting $g + Q * f$ for f on the right side of (2.42) repeatedly we obtain

$$f = \sum_{r=0}^{n} Q^r * g + Q^{n+1} * f \quad (n \geq 0). \tag{2.43}$$

Since $g \geq 0$, the sum on the right side of (2.43) increases as $n \to \infty$ to $U * g$, while $Q^{n+1} * f$ tends to some function h. This gives $f = U * g + h$. From

$$(Q^{n+1} * f)_j(t) = \sum_{l \in E} \int_{0-}^{t} Q_{jl}\{ds\} (Q^n * f)_l(t-s)$$

we see that h satisfies (2.41). □

Theorem 2.6. *The solution of the Markov renewal equation is unique iff $N(t) < \infty$ for all $t \in \mathbf{R}_+$ a.s. In particular, if E is finite, the solution is unique.*

Proof. We have seen that $q_j(t) = P_j\{N(t) = \infty\}$ satisfies the integral equation (2.24), which is essentially the equation (2.41) with solutions restricted to $0 \leq h_j(t) \leq 1$. It can be proved that this restriction does not result in any loss of generality. Thus, in view of Theorem 2.2, the only solution to (2.41) is $h_j(t) \equiv 0$ iff $N(t) < \infty$ for all $t \in \mathbf{R}_+$. In this case (2.39) is the unique solution of the Markov renewal equation. The case of finite E follows directly from the proof of Theorem 2.5 since in (2.43)

$$\lim_{n\to\infty} (Q^{n+1} * f)_j(t) = \sum_{l\in E} \int_{0-}^{t} \lim_{n\to\infty} Q_{jl}^{(n+1)}\{ds\} f_l(t-s) = 0$$

on account of the fact that

$$\lim_{n\to\infty} Q_{jl}^{(n+1)}\{[0,t]\} = 0. \qquad \square$$

2.6 Limit Theorems

We now investigate the limit behavior of the renewal function $U_{jk}(t)$ and the solution (2.39) of the Markov renewal equation. The following theorems are consequences of standard renewal theory.

Theorem 2.7. *(a) If the imbedded renewal process S^k is terminating, then*

$$\lim_{t\to\infty} U_{jk}(t) = F_{jk} U_{kk}(\infty). \tag{2.44}$$

(b) If S^k is nonterminating, then

$$\lim_{t\to\infty} \frac{U_{jk}(t)}{t} = \frac{F_{jk}}{\mu_{kk}}. \tag{2.45}$$

(c) If S^k is nonterminating and continuous, then as $t \to \infty$

$$U_{jk}(t) - U_{jk}(t-h) \to \frac{hF_{jk}}{\mu_{kk}} \quad (h > 0). \tag{2.46}$$

Here $F_{jk} = P_j\{S_1^k < \infty\}$, $\mu_{kk} = E_k(S_1^k)$ and the limits in (2.45)-(2.46) are interpreted as zero if $\mu_{kk} = \infty$.

Theorem 2.8. *Let E be finite. If in (2.39) the functions $g_j(t)$, $j \in E$ are directly Riemann integrable, then*

$$\lim_{t\to\infty} G_j(t) = \sum_{l\in E} \frac{F_{jl}}{\mu_{ll}} \int_0^\infty g_l(s)\,ds. \tag{2.47}$$

2.7 The Semi-Markov Process

We define a process $J = \{J(t), t \geq 0\}$ as follows. Let $J(0) = J_0$ and

$$J(t) = \begin{cases} J_n & \text{for } S_n \leq t < S_{n+1} \quad (n \geq 0) \\ \Delta & \text{for } t \geq L \end{cases} \tag{2.48}$$

where Δ is a point of compactification of E and $L = \sup_{n\geq 0} S_n$, as defined by (2.31). The process J is called the minimal semi-Markov process associated with the MRP $\{(S_n, J_n), n \geq 0\}$. Denote

$$P_{jk}(t) = P\{J(t) = k | J(0) = j\}, \quad j, k \in E,\ t \geq 0. \tag{2.49}$$

Theorem 2.9. *We have*

$$P_{jk}(t) = \int_{0-}^{t} U_{jk}\{ds\} P_k\{S_1 > t - s\}. \tag{2.50}$$

Proof. An easy calculation shows that

$$P_{jk}(t) = P_j\{S_1 > t\}\, \delta_{jk} + \sum_{l\in E} \int_{0-}^{t} Q_{jl}\{ds\} P_{lk}(t-s). \tag{2.51}$$

This shows that for fixed $k \in E$, $P_{jk}(t)$ satisfies the Markov renewal equation (2.35) with

$$g_j(t) = P_j\{S_1 > t\}\, \delta_{jk}.$$

The minimal solution of (2.51) is given by

$$\sum_{l\in E} \int_{0-}^{t} U_{jl}\{ds\} g_l(t-s) = \int_{0-}^{t} U_{jk}\{ds\} P_k\{S_1 > t - s\}.$$

By Theorem 2.5 we therefore obtain

$$P_{jk}(t) \geq \int_{0-}^{t} U_{jk}\{ds\} P_k\{S_1 > t - s\}. \tag{2.52}$$

If strict inequality holds in (2.52) for some $k \in E$, then

$$\begin{aligned} P_j\{L > t\} &= \sum_{l\in E} P_{jk}(t) \\ &> \sum_{k\in E} \int_{0-}^{t} U_{jk}\{ds\} \left[1 - \sum_{l\in E} Q_{kl}\{[0, t-s]\}\right] \\ &= \lim_{n\to\infty} \sum_{m=0}^{n} \sum_{k\in E} \int_{0-}^{t} Q_{jk}^{(m)}\{ds\} \left[1 - \sum_{l\in E} Q_{kl}\{[0, t-s]\}\right] \\ &= \lim_{n\to\infty} \sum_{m=0}^{n} \left[\sum_{k\in E} Q_{jk}^{(m)}\{[0,t]\} - \sum_{l\in E} Q_{jl}^{(m+1)}\{[0,t]\}\right] \\ &= \lim_{n\to\infty} [1 - q_j^{(n+1)}(t)] \end{aligned}$$

in the notation of Sec. 2.4. Since

$$\lim_{n\to\infty} q_j^{(n+1)}(t) = q_j(t) = P_j\{N(t) = \infty\} = P_j\{L \le t\}$$

the equality must hold in (2.52) for every $k \in E$, and (2.50) is the desired solution of the equation (2.51). □

The following result follows from (2.50) and the key renewal theorem.

Theorem 2.10. *If the renewal process S^k is nonterminating and continuous, with $\mu_{kk} = E_k(S_1^k) < \infty$, then as $t \to \infty$,*

$$P_{jk}(t) \to \frac{F_{jk}}{\mu_{kk}} E_k(X_1). \tag{2.53}$$

Example 2.3. (The pure birth semi-Markov process). This is the J-process associated with the MRP of Example 2.2. Suppose $J(0) = j$. Then

$$J(t) = \begin{cases} j + n & \text{for } S_n \le t < S_{n+1}\ (n \ge 0) \\ \Delta & \text{for } t \ge L. \end{cases}$$

Here $L < \infty$ a.s., as we have already seen.

If the distribution F_j has the exponential density $\lambda_j e^{-\lambda_j x}(0 < \lambda_j < \infty)$, then J reduces to the pure birth process.

2.8 Markov-Additive Processes: Basic Definitions

We are given a probability space $(\Omega, \mathcal{F}, P)$ and denote $\mathbf{R} = (-\infty, \infty)$, $E =$ a countable set and $\mathbf{N}_+ = \{0, 1, 2, \ldots\}$.

Definition 2.2. A Markov-additive process $(X, J) = \{(X(t), J(t)), t \ge 0\}$ is a two-dimensional Markov process on the state space $\mathbf{R} \times E$ such that, for $s, t \ge 0$, the conditional distribution of $(X(s+t) - X(s), J(s+t))$ given $(X(s), J(s))$ depends only on $J(s)$.

Since (X, J) is Markov, it follows easily from the definition that J is Markov and that X has conditionally independent increments, given the states of J, i.e., for $0 \le t_1 \le \ldots \le t_n$ $(n \ge 2)$ the increments

$$X(t_1) - X(0),\ X(t_2) - X(t_1), \ldots,\ X(t_n) - X(t_{n-1})$$

are conditionally independent given $J(0), J(t_1), \ldots, J(t_n)$. Since in general X is non-Markovian, we may therefore call J the Markov component and X the additive component of the MAP (X, J). Note that when the state space of J consists of only one element, X has independent increments, thus implying that a Markov-additive process is a generalization of Lévy process, but in general the additive component of an MAP does not have additive increments.

Definition 2.2 is in the spirit of Çinlar [28] who considers more general state space for J. A second definition of MAPs is given by Çinlar [29], who starts with the Markov component J and defines X as a process having properties relative to J that imply, in particular, the conditions stated in Definition 2.2. In the discrete time analog, this amounts to viewing the additive component X as sums of random variables defined on a Markov chain, rather than formulating (X, J) as a MRW (Markov random walk) in the way followed in Chap. 4. In such an approach X bears a causal relationship with J, which is assumed to be more or less known at the beginning. This is the case in many applications where J represents an extraneous factor such as the environment (e.g., in some Markov-modulated queueing systems, as illustrated in chapters 7–9). However, in other applications, the phenomenon studied gives rise in a natural fashion to X and J jointly, and it is important to study the evolution of (X, J) as a Markov process. In such situations, Definition 2.2 is a natural one to use.

In the chapter we address only the homogeneous case, for which the conditional distribution of $(X(s+t) - X(s), J(s+t))$ given $J(s)$ depends only on t. In this case, it follows that

$$\begin{aligned} P\{X(s+t) \in A, J(s+t) = k | X(s) = x, J(s) = j\} \\ = P\{X(s+t) - X(s) \in A - x, J(s+t) = k | J(s) = j\} \\ = P\{X(t) - X(0) \in A - x, J(t) = k | J(0) = j\} \end{aligned}$$

for $j, k \in E$, $x \in \mathbf{R}$ and a Borel subset A of $\mathbf{R}$. Therefore it suffices to define the transition distribution measure of the process as

$$F_{jk}(A; t) = P\{X(t) \in A, J(t) = k | J(0) = j\} \tag{2.54}$$

$$F_{jk}(A; t) \geq 0, \sum_{k \in E} F_{jk}(\mathbf{R}; t) = 1 \tag{2.55}$$

$$F_{jk}(A; 0) = \begin{cases} 1 & \text{if } j = k, 0 \in A \\ 0 & \text{otherwise.} \end{cases} \tag{2.56}$$

For the Chapman-Kolmogorov equations we have

$$F_{jk}(A; t+s) = \sum_{l \in E} \int_{\mathbf{R}} F_{jl}(dx; t) F_{lk}(A - x; s) \quad (s, t \geq 0). \tag{2.57}$$

From (2.54) the transition probabilities of J are found to be

$$\pi_{jk}(t) = P\{J(t) = k|J(0) = j\} = F_{jk}(\mathbf{R}; t). \tag{2.58}$$

The $\pi_{jk}(t)$ satisfy conditions similar to (2.55-2.57). Thus

$$\pi_{jk}(t) \geq 0, \quad \sum_{k \in E} \pi_{jk}(t) = 1, \quad \pi_{jk}(0) = \delta_{jk} \tag{2.59}$$

and

$$\pi_{jk}(t+s) = \sum_{l \in E} \pi_{jl}(t)\pi_{lk}(s). \tag{2.60}$$

Further characterization of the process will depend on the properties of the increments $X(t) - X(0)$. We shall only consider the case where all the states of J are stable,with transition rates $\nu_{jk}(j \neq k)$. We denote $\nu_{jj} = -\sum_{k \neq j} \nu_{jk}(-\infty < \nu_{jj} < 0)$ and $N = (\nu_{jk})$, so that N is the generator matrix of J. Although most of our derivations remains valid for a countable set E, in order to get explicit results we shall assume in the rest of the chapter that E is finite, say $E = \{1, 2, \ldots, N\}$. In this case N is a finite matrix of order N, and it is known (and will also follow from our results) that

$$\pi(t) = (\pi_{jk}(t)) = e^{tN}. \tag{2.61}$$

It turns out that a similar result holds for the two-dimensional process (X, J). The clue is provided by taking the transforms

$$F^*_{jk}(w; t) = \int_{\mathbf{R}} e^{iwx} F_{jk}(dx; t) \tag{2.62}$$

where w is real and $i = \sqrt{-1}$. The Chapman-Kolmogorov equations (2.57) then yield the relations

$$F^*_{jk}(w; t+s) = \sum_{l \in E} F^*_{jl}(w; t) F^*_{lk}(w; s) \tag{2.63}$$

or in the matrix form, writing $F^*(w; t) = (F^*_{jk}(w; t))$,

$$F^*(w; t+s) = F^*(w; t) F^*(w; s). \tag{2.64}$$

We seek a solution of (2.64) in the form (2.61), namely,

$$F^*(w; t) = e^{tQ(w)} \qquad (t \geq 0). \tag{2.65}$$

We establish this solution in special cases. In order to do so, we make some assumptions concerning the infinitesimal transitions of the process, namely concerning $F_{jk}(A; h)$ for h small. We then derive a system of differential equations for $F_{jk}(A;h)$ using (2.57). In terms of the transforms (2.62), this system can by written as

$$\frac{\partial}{\partial t}F^*(w;t) = F^*(w;t)Q(w). \tag{2.66}$$

Our assumptions will guarantee the continuity of $F^*_{jk}(w;t)$ as a function of t. For finite E it will be found that (2.66) has the unique solution (2.65) as desired. A preliminary result to this effect is proved in the next section. In the general case of a countable set E the use of infinitesimal generators will be explained.

2.9 A Matrix Equation

We consider the matrix equation

$$P'(t) = P(t)Q \tag{2.67}$$

where P is a matrix of continuous (real or complex) functions of t and Q is a matrix of constants (that is, elements not depending on t), both P and Q being square matrices of finite order n. In order to solve this equation we need the concept of the norm of a matrix. For a square matrix $A = (a_{jk})$ of finite order n, we define the norm $\|A\|$ as

$$\|A\| = \sum_{j=1}^{n}\sum_{k=1}^{n}|a_{jk}|. \tag{2.68}$$

This norm has the following properties:

(a) $\|A\| = 0$ iff $A = 0$.
(b) If c is a constant, then $\|cA\| = |c|\,\|A\|$.
(c) If A and B are square matrices of order n, then

$$\|A+B\| \leq \|A\| + \|B\|, \quad \|AB\| \leq \|A\|\|B\|.$$

(d) If the elements $a_{jk}(t)$ of $A(t)$ are continuous functions of t, then we define

$$\int_{c_1}^{c_2} A(t)\,dt = \left(\int_{c_1}^{c_2} a_{jk}(t)\,dt\right) \quad (c_1 < c_2).$$

We have then

$$\left\|\int_{c_1}^{c_2} A(t)\,dt\right\| \leq \int_{c_1}^{c_2} \|A(t)\|\,dt.$$

Theorem 2.11. *The unique solution of (2.67) subject to the condition $P(0) = I$ is given by*

$$P(t) = e^{tQ}. \tag{2.69}$$

Proof. *(i).* Let $T \geq 0$ be arbitrary and $m = \|Q\| > 0$. We prove that (2.67) has at most one solution. Suppose P_1 and P_2 are two solutions and

$$m_1 = \max_{0 \leq t \leq T} \|P_1(t) - P_2(t)\|, \quad 0 < m_1 < \infty. \tag{2.70}$$

Then

$$P_1(t) - P_2(t) = \int_0^t [P_1(s) - P_2(s)]\, Q\, ds \tag{2.71}$$

and so, for $0 \leq t \leq T$,

$$\|P_1(t) - P_2(t)\| \leq \int_0^t \|P_1 - P_2\|\, \|Q\|\, ds \leq m_1 m\, t. \tag{2.72}$$

Using (2.72) in (2.71) we obtain

$$\|P_1(t) - P_2(t)\| \leq \int_0^t m_1\, m\, s\, \|Q\|\, ds = m_1 \frac{(mt)^2}{2!}$$

and by induction

$$\|P_1(t) - P_2(t)\| \leq m_1 \frac{(mt)^n}{n!} \longrightarrow 0 \text{ as } n \to \infty. \tag{2.73}$$

It follows that $\|P_1(t) - P_2(t)\| = 0$, $0 \leq t \leq T$, and therefore $P_1 = P_2$, as desired.

(ii). For $P(t)$ given by (2.69) we have

$$\frac{P(t+h) - P(t)}{h} = \frac{e^{(t+h)Q} - e^{tQ}}{h} = P(t)\frac{e^{hQ} - I}{h}$$

where

$$\left\| \frac{e^{hQ} - I}{h} - Q \right\| = \left\| \sum_{n=2}^{\infty} \frac{h^{n-1}}{n!} Q^n \right\| \leq \sum_{n=2}^{\infty} \frac{h^{n-1} m^n}{n!} \to 0$$

as $h \to 0^+$. This means that

$$\frac{e^{hQ} - I}{h} \to Q \text{ as } h \to 0_+$$

and $P'(t) = P(t)Q$. Therefore $P(t)$ given by (2.69) is a solution of (2.67), and by *(i)* it is the unique solution. □

2.10 The Markov-Poisson Process

This is a Markov-additive process (X, J) on the state space $\mathbf{N}_+ \times E$, such that X is a simple Cox process (doubly stochastic Poisson process) whose rate depends on the state of J. This process is also known in the literature as MMPP (Markov-modulated Poisson process) and belongs to the class of MAPs of arrivals, studied in Chap. 6. Specifically, denote the transition probabilities of the process by

$$P_{jk}(n;t) = P\{X(t) = n, J(t) = k | J(0) = j\} \quad j, k \in E, n \in \mathbf{N}_+. \tag{2.74}$$

Then, in addition to the assumptions of Definition 2.2, we assume that

$$P_{jk}(n;h) = a_{jk}(n)h + o(h) \tag{2.75}$$

where the transition rates $a_{jk}(n)$ are given by

$$a_{jj}(1) = \lambda_j, \quad a_{jk}(0) = \nu_{jk} \quad (k \neq j), \tag{2.76}$$

all other $a_{jk}(n)$ being zero. Here $0 \leq \lambda_j < \infty$. It should be noted that during a time-interval $(t, t+h]$ with $h \to 0^+$, changes of state occur either in X or J, but not both simultaneously; thus second order effects are ignored. The Chapman-Kolmogorov equations for this process are

$$P_{jk}(n;t+s) = \sum_{l \in E} \sum_{m=0}^{n} P_{jl}(m;t) P_{lk}(n-m;s) \quad (s, t \geq 0). \tag{2.77}$$

Considering the process over the time-intervals $(0, t]$, $(t, t+dt]$ and using (2.77) we find that

$$\begin{aligned} P_{jk}(n;t+dt) &= P_{jk}(n;t)(1 - \lambda_k dt + \nu_{kk} dt) \\ &\quad + P_{jk}(n-1;t)\lambda_k dt + \sum_{l \neq k} P_{jl}(n;t)\nu_{lk} dt + o(dt). \end{aligned}$$

This leads to the system of differential equations

$$\begin{aligned} \frac{\partial}{\partial t} P_{jk}(n;t) &= -\lambda_k P_{jk}(n;t) + \lambda_k P_{jk}(n-1;t) \\ &\quad + \sum_{l \in E} P_{jl}(n;t)\nu_{lk}. \end{aligned} \tag{2.78}$$

To solve this we introduce the generating functions

$$G_{jk}(z,t) = \sum_{n=0}^{\infty} P_{jk}(n;t) z^n \quad (0 < z < 1). \tag{2.79}$$

Then (2.78) reduces to

$$\frac{\partial}{\partial t}G_{jk}(z,t) = -\lambda_k(1-z)G_{jk}(z,t) + \sum_{l\in E} G_{jl}(z,t)\nu_{lk} \tag{2.80}$$

for $j,k \in E$. These equations may be expressed in the matrix from as follows. Let

$$G(z,t) = (G_{jk}(z,t)), \qquad \Lambda = (\lambda_j\delta_{jk}) \tag{2.81}$$

and recall that $N = (\nu_{jk})$ is the generator matrix of J. Then

$$\frac{\partial}{\partial t}G(z,t) = G(z,t)(-\Lambda + \Lambda z + N). \tag{2.82}$$

For finite E, Theorem 2.11 yields the following.

Theorem 2.12. *The matrix of generating functions of the Markov-Poisson process defined by (2.75)-(2.76) is given by*

$$G(z,t) = e^{-t(\Lambda-\Lambda z-N)}. \tag{2.83}$$

Letting $z \to 1$ in (2.83) we find that

$$\pi(t) = G(1^-,t) = e^{tN} \tag{2.84}$$

in agreement with (2.61).

We may view our process as describing the occurrence of an event E such that the rate of occurrence λ is modulated by the underlying Markov process J, and hence a random variable denoted by $\lambda_{J(t)}$ at time t. The following properties of this process are then direct extension of those in the standard Poisson process.

(a) The means. For each $t \geq 0$ the distribution of $\{X(t), J(t)\}$ is proper, since by (2.84) and (2.59)

$$P\{X(t) \in \mathbf{N}_+, J(t) \in E | J(0) = j\} = G(1-,t)\mathbf{e} = \pi(t)\mathbf{e} = 1 \tag{2.85}$$

where $\mathbf{e}$ is the column vector with unit elements. Now let

$$e_{jk}(t) = E[X(t); J(t) = k | J(0) = j] = \sum_{n=0}^{\infty} nP_{jk}(n;t). \tag{2.86}$$

From (2.78) we find that the $e_{jk}(t)$ satisfy the differential equations

$$e'_{jk}(t) = \sum_{l\in E} e_{jl}(t)\nu_{lk} + \pi_{jk}(t)\lambda_k, \quad k \in E. \tag{2.87}$$

Adding these over $k \in E$ we obtain

$$\frac{d}{dt}E[X(t)|J(0) = j] = \sum_{k\in E} \pi_{jk}(t)\lambda_k.$$

This gives

$$E[X(t)|J(0)=j] = \sum_{k\in E} \int_0^t \pi_{jk}(s)\lambda_k ds \tag{2.88}$$

since $e_{jk}(0) = 0$. If the chain J has a limit distribution $\{\pi_k, k \in E\}$, then (2.88) gives

$$\lim_{t\to\infty} t^{-1}E[X(t)|J(0)=j] = \sum_{k\in E} \pi_k \lambda_k. \tag{2.89}$$

Returning to $e_{jk}(t)$ we claim that the unique solutions of (2.87) is given by

$$e_{jk}(t) = \sum_{l\in E} \int_0^t \pi_{jl}(s)\lambda_l \pi_{lk}(t-s)\, ds.$$

(b) Lack of memory property. Let T be the epoch of first occurrence of the event E, so that

$$T = \inf\{t : X(t) = 1\}. \tag{2.90}$$

We have

$$P\{T > t, J(t) = k|J(0) = j\} = P_{jk}(0;t). \tag{2.91}$$

From (2.77) we find that

$$P_{jk}(0;t+s) = \sum_{l\in E} P_{jl}(0;t)P_{lk}(0;s)$$

which can be written as

$$\begin{aligned} &P\{T > t+s, J(t+s) = k|J(0) = j\} \\ &= \sum_{l\in E} P\{T > t, J(t) = l|J(0) = j\}P\{T > s, J(s) = k|J(0) = l\}. \end{aligned} \tag{2.92}$$

This relation characterizes the lack of memory property in the Markov-modulated case.

(c) Interocurrence times. The epochs of successive occurrences of events are given by T_r, where

$$T_r = \inf\{t : X(t) = r\} \qquad (r \geq 1) \tag{2.93}$$

where $T_1 = T$, as defined by (2.90). We shall also denote $T_0 = 0$. Owing to the presence of J we expect the interocurrence times $T_r - T_{r-1}$ $(r \geq 1)$ to be Markov-dependent. The following result makes this statement more precise. We assume that the process (X, J) is a strong Markov process.

Theorem 2.13. *Let T_r be defined by (2.93) and denote $J_r = J \circ T_r$ $(r \geq 0)$. Then $\{(T_r, J_r), r \geq 0\}$ is an MRP whose transition density is given by*

$$(q_{jk}(t)) = e^{-t(\Lambda - N)}\Lambda. \tag{2.94}$$

Moreover, this density is a proper one, with mean

$$\left(\sum_{k \in E} \int_0^\infty t\, q_{jk}(t)\, dt \right) = (\Lambda - N)^{-1}\mathbf{e}. \tag{2.95}$$

Proof. From

$$T_{r+1} - T_r = \inf\{t - T_r : X(t) - X(T_r) = 1\}$$

we see that given $J_0, T_1, J_1, \ldots, T_r, J_r$, the distribution of $(T_{r+1} - T_r, J_{r+1})$ is the same as that of (T_1, J_1), given J_0. This leads to the Markov renewal property. For the transition distribution measure of this process we have

$$Q_{jk}\{dt\} = P_{jk}(0; t)\lambda_k dt = q_{jk}(t)\, dt$$

where the transition density $q_{jk}(t)$ is given by

$$q_{jk}(t) = P_{jk}(0, t)\lambda_k.$$

This gives

$$(q_{jk}(t)) = (P_{jk}(0, t))\Lambda$$

which is the desired result (2.94), since from Theorem 2.12

$$(P_{jk}(0, t)) = G(0^+, t) = e^{-t(\Lambda - N)}.$$

To show that the distribution Q_{jk} is proper, we note from (2.94) that

$$\sum_{k \in E} \int_{0^-}^\infty q_{jk}(t)\, dt = (\Lambda - N)^{-1}\Lambda\mathbf{e} = (\Lambda - N)^{-1}(\Lambda - N)\mathbf{e} = 1$$

since $N\mathbf{e} = 0$. Again, from (2.94)

$$\begin{aligned} \sum_{k \in E} \int_0^\infty t\, q_{jk}(t)\, dt &= (\Lambda - N)^{-2}\Lambda\mathbf{e} = (\Lambda - N)^{-2}(\Lambda - N)\mathbf{e} \\ &= (\Lambda - N)^{-1}\mathbf{e}. \end{aligned}$$

□

2.11 Markov-Compound Poisson Processes; A Special Case

Suppose that $\{A(t), t \geq 0\}$ and $\{D(t), t \geq 0\}$ are simple Markov-Poisson processes which are jointly modulated by an underlying Markov process J as in Sec. 2.10. We consider the process (X, J), with $X(t) = A(t) - D(t)$, which is clearly a Markov-additive process on the state space $\mathbf{N} \times E$, where $\mathbf{N} = \{\ldots, -1, 0, 1, 2, \ldots\}$. Here the jumps in $X(t)$ of size +1 or -1 occur at rates depending on the state of J. This process belongs to the class of Markov-compound Poisson processes, which will be considered in the next section.

Let us denote the transition probabilities of the process by

$$P_{jk}(n;t) = P\{X(t) = n, J(t) = k | J(0) = j\} \quad j, k \in E, n \in \mathbf{N}_+. \tag{2.96}$$

The transition rates (2.75) are given in this case by

$$a_{jj}(1) = \lambda_j, \quad a_{jj}(-1) = \mu_j, \quad a_{jk}(0) = \nu_{jk} \quad (j \neq k) \tag{2.97}$$

all other $a_{jk}(n)$ being zero. Here λ_k and μ_k are respectively the transition rates associated with the processes $A(t)$ and $D(t)$. Proceeding as in Sec. 2.10 we obtain the system of differential equations

$$\begin{aligned}\frac{\partial}{\partial t} P_{jk}(n;t) = &-(\lambda_k + \mu_k) P_{jk}(n;t) + P_{jk}(n-1;t)\lambda_k \\ &+ P_{jk}(n+1;t)\mu_k + \sum_{l \in E} P_{jl}(n;t)\nu_{lk}.\end{aligned} \tag{2.98}$$

We now introduce the generating functions

$$G_{jk}(z,t) = \sum_{n \in E} P_{jk}(n;t) z^n \quad (|z| \leq 1). \tag{2.99}$$

Then (2.98) reduces to

$$\begin{aligned}\frac{\partial}{\partial t} G_{jk}(z,t) = &-(\lambda_k - \lambda_k z + \mu_k - \mu_k z^{-1}) G_{jk}(z,t) \\ &+ \sum_{l \in E} G_{jl}(z,t)\nu_{lk}\end{aligned}$$

or in the matrix form

$$\frac{\partial}{\partial t} G(z,t) = G(z,t)(-\Lambda + \Lambda z - M + M z^{-1} + N) \tag{2.100}$$

where the matrices $G(z,t)$, Λ, and N are as in Sec. 2.10 and

$$M = (\mu_j \delta_{jk}). \tag{2.101}$$

For finite E, Theorem 2.11 yields the following.

Theorem 2.14. *The matrix of generating functions of the process (X, J) defined by (2.96) and (2.97) is given by*

$$G(z,t) = e^{-t(\Lambda - \Lambda z + M - M z^{-1} - N)}. \tag{2.102}$$

We now consider the random variables

$$T_r = \inf\{t : X(t) = r\} \tag{2.103}$$

where $T_0 = 0$ and for $r \neq 0$, T_r is the hitting time of the level $X(t)$, or more appropriately, the hitting time of the set $\{r\} \times E$ of the process (X, J). It suffices to consider the case $r > 0$, since the case $r < 0$ can obtained by an interchange of the matrices Λ and M. We have the following.

Theorem 2.15. *For $r \geq 0$ let the random variables T_r be defined by (2.103) and denote $J_r = J \circ T_r$. Then $\{(T_r, J_r), r \geq 0\}$ is an MRP. Its transition distribution measure has the transform*

$$\xi_{jk}(s) = E[e^{-sT_1}; J_1 = k | J_0 = j] \tag{2.104}$$

where the matrix $\xi(s) = (\xi_{jk}(s))$ satisfies the equation

$$M\xi^2 - (sI + \Lambda + M - N)\xi + \Lambda = 0. \tag{2.105}$$

Proof. The Markov renewal property follows as in the proof of Theorem 2.13. Let us denote the transition measures of the MRP as

$$Q_{jk}^{(r)}\{A\} = P\{T_r \in A, J_r = k | J_0 = j\}$$

and their transforms as

$$\hat{Q}_{jk}^{(r)}(s) = \int_{0-}^{\infty} e^{-st} Q_{jk}^{(r)}\{dt\} \qquad (s > 0)$$

for $r \geq 0$. Here $\hat{Q}_{jk}^{(1)}(s) = \xi_{jk}(s)$ as defined by (2.104). It is known that

$$(\hat{Q}_{jk}^{(r)}(s)) = \xi(s)^r. \tag{2.106}$$

We need to prove that $\xi(s)$ satisfies (2.105). Considering the epoch of the first jump in (X, J) we find that for $r \geq 1$

$$\begin{aligned} Q_{jk}^{(r)}(t) = &\int_0^t e^{-(\lambda_j + \mu_j - \nu_{jj})\tau} \lambda_j Q_{jk}^{(r-1)}(t - \tau) d\tau \\ &+ \int_0^t e^{-(\lambda_j + \mu_j - \nu_{jj})\tau} \mu_j Q_{jk}^{(r+1)}(t - \tau) d\tau \\ &+ \sum_{l \neq j} \int_0^t e^{-(\lambda_j + \mu_j - \nu_{jj})\tau} \nu_{jl} Q_{lk}^{(r)}(t - \tau) d\tau, \end{aligned}$$

where we have written $Q_{jk}^{(r)}(t)$ for $Q_{jk}^{(r)}\{[0, t]\}$ for convenience. Taking transforms of both sides of this last equation, we obtain

$$(s + \lambda_j + \mu_j - \nu_{jj})\hat{Q}_{jk}^{(r)}(s) = \lambda_j \hat{Q}_{jk}^{(r-1)}(s) + \mu_j \hat{Q}_{jk}^{(r+1)}(s) + \sum_{l \neq j} \nu_{jl} \hat{Q}_{lk}^{(r)}(s)$$

or in the matrix form, using (2.106),

$$(sI + \Lambda + M - N)\xi^r = \Lambda \xi^{r-1} + M \xi^{r+1}.$$

This gives (2.105) as desired.

The existence and uniqueness of the solution of the equation (2.105) involves the use of matrix algebra and will not be discussed here. □

2.12 Markov-Compound Poisson Processes; General Case

We define a Markov-compound Poisson process $(X, J) = \{(X(t), J(t)), t \geq 0\}$ on the state space $\mathbf{R} \times E$ as follows. Changes in state of the additive component X occur continuously because of a drift $d_{J(t)}$ modulated by the Markov process J, and by jumps that are attributed either to the Poisson process or else to J. Let us denote by B_j the distribution of the Poisson jumps occurring at a rate λ_j when J is in state j, and by B_{jk} the distribution of the Markov-modulated jumps occurring when J changes from state j to state k, which it does at a rate $\nu_{jk} (k \neq j)$. Here the distributions B_j and B_{jk} are defined over $\mathbf{R}$, so that positive as well as negative jumps occur. The probability of no jump during a time-interval $(t, t+h]$ if at time t the chain J is in state j is $1 - \lambda_j h + \nu_{jj} h + o(h)$. Our assumptions may be summarized as follows. Denote by

$$F_{jk}(x; t) = P\{X(t) \leq x, J(t) = k | J(0) = j\} \tag{2.107}$$

the transition distribution function of the process. Then

$$\begin{aligned} F_{jk}(x; h) = \; & \delta_{jk}[1 - \lambda_j h + \nu_{jj} h + o(h)]\varepsilon_0(x - d_j h) \\ & + \delta_{jk}[\lambda_j h B_j(x - d_j h) + o(h)] \\ & + (1 - \delta_{jk})[\nu_{jk} h B_{jk}(x - d_j h) + o(h)] \end{aligned} \tag{2.108}$$

where the distribution ε_0 is concentrated at the origin. Considering (X, J) over the time-intervals $(0, t], (t, t+h]$ and using the Chapman-Kolmogorov equations (2.57) we find that

$$\begin{aligned} F_{jk}(x; t+h) = \; & F_{jk}(x - d_k h; t)(1 - \lambda_k h + \nu_{kk} h) \\ & + \int_{\mathbf{R}} F_{jk}(x - v - d_k h; t) \lambda_k h B_k\{dv\} \\ & + \sum_{l \neq k} \int_{\mathbf{R}} F_{jl}(x - v - d_l h; t) \nu_{lk} h B_{lk}\{dv\} + o(h). \end{aligned}$$

This leads to the system of integro-differential equations

$$\begin{aligned} & \frac{\partial}{\partial t} F_{jk}(x; t) + d_k \frac{\partial}{\partial x} F_{jk}(x; t) \\ & \qquad = -\lambda_k F_{jk}(x; t) + \int_{\mathbf{R}} F_{jk}(x - v; t) \lambda_k B_k\{dv\} \\ & \qquad\quad + \sum_{l \in E} \int_{\mathbf{R}} F_{jl}(x - v; t) \nu_{lk} B_{lk}\{dv\} \end{aligned} \tag{2.109}$$

where $B_{kk} = \varepsilon_0$. To solve (2.109) we introduce the transforms

$$\psi_j(w) = \int_{\mathbf{R}} e^{iwx} B_j\{dx\}, \quad \psi_{jk}(w) = \int_{\mathbf{R}} e^{iwx} B_{jk}\{dx\} \tag{2.110}$$

and recall that $F^*(w;t)$ is defined by (2.62). Also, denote the matrices

$$D = (d_j \delta_{jk}), \quad \Psi(w) = (\psi_j(w)\delta_{jk}), \quad N(w) = (\nu_{jk}\psi_{jk}(w)) \tag{2.111}$$

and $F^*(w;t) = (F^*_{jk}(w;t))$. Note that $N(0) = N$, the generator matrix of J. For finite E we have the following result.

Theorem 2.16. *The solution of the system of equations (2.109) is given by*

$$F^*(w;t) = e^{tQ(w)} \tag{2.112}$$

where

$$Q(w) = iwD - \Lambda + \Lambda\Psi(w) + N(w). \tag{2.113}$$

Proof. We make the transformation $(x,t) \to (y,t)$ with $y = x - d_k t$. The function $F_{jk}(x;t)$ is then transformed into $G_{jk}(y;t)$ and so (with obvious notations)

$$F^*_{jk}(w;t) = \int_{\mathbf{R}} e^{iw(y+d_k t)} G_{jk}(dy;t) = e^{iwd_k t} G^*_{jk}(w;t)$$

or in the matrix form

$$F^*(w;t) = G^*(w;t)e^{iwtD}. \tag{2.114}$$

Now we have

$$\frac{\partial}{\partial x}F_{jk} = \frac{\partial}{\partial y}G_{jk}\frac{dy}{dx} + \frac{\partial}{\partial t}G_{jk}\frac{dt}{dx} = \frac{\partial}{\partial y}G_{jk}$$

$$\frac{\partial}{\partial t}F_{jk} = \frac{\partial}{\partial y}G_{jk}\frac{dy}{dt} + \frac{\partial}{\partial t}G_{jk} = -d_k\frac{\partial}{\partial y}G_{jk} + \frac{\partial}{\partial t}G_{jk}$$

so that

$$\frac{\partial}{\partial t}F_{jk} + d_k\frac{\partial}{\partial x}F_{jk} = \frac{\partial}{\partial t}G_{jk} \tag{2.115}$$

and the equations (2.109) reduce to

$$\frac{\partial}{\partial t}G_{jk}(y;t) = -\lambda_k G_{jk}(y;t) + \int_{\mathbf{R}} G_{jk}(y-v;t)\lambda_k B_k\{dv\}$$
$$+ \sum_{l \in E}\int_{\mathbf{R}} G_{jl}(y-v;t)\nu_{lk}B_{lk}\{dv\}. \tag{2.116}$$

Taking transforms of both sides of (2.116) we obtain

$$\frac{\partial}{\partial t}G^*_{jk}(w;t) = -\lambda_k G^*_{jk}(w;t) + \lambda_k G^*_{jk}(w;t)\psi_k(w) + \sum_{l\in E} G^*_{jl}(w;t)\nu_{lk}\psi_{lk}(w)$$

or in the matrix form

$$\frac{\partial}{\partial t}G^*(w;t) = G^*(w;t)[-\Lambda + \Lambda\Psi(w) + N(w)]. \tag{2.117}$$

From Theorem 2.11 we obtain the solution of (2.117) as

$$G^*(w;t) = e^{t[-\Lambda+\Lambda\Psi(w)+N(w)]}. \tag{2.118}$$

The desired result now follows from (2.114) and (2.118). □

2.13 The Use of Infinitesimal Generators

For the general case of Markov-additive processes we illustrate the use of infinitesimal generators as follows. We assume

$$F_{jk}(dx;h) = \delta_{jk}(1+\nu_{jj}h)H_j(dx;h) + (1-\delta_{jk})\nu_{jk}hB_{jk}\{dx\} + o(h) \tag{2.119}$$

where $H_j(A;t)$ is the conditional distribution measure (given that J is in state j during t units of time) of a Lévy process and B_{jk} is the distribution of Markov-modulated jumps as in Sec. 2.12. Let $f(x,j)$ be a bounded function on $\mathbf{R}\times E$, such that for each fixed j, f is continuous and has a bounded continuous derivative $\partial f/\partial t$. The infinitesimal generator of the process is defined as the operator $\mathcal{A}$ where

$$\mathcal{A}f(x,j) = \lim_{h\to 0^+} h^{-1}\sum_{k\in E}\int_{\mathbf{R}}[f(x+y,k)-f(x,j)]F_{jk}(dy;h). \tag{2.120}$$

The following result follows immediately from (2.119).

Theorem 2.17. *The infinitesimal generator of the Markov-additive process defined by (2.119) is given by $\mathcal{A}$, where*

$$\mathcal{A}f(x,j) = \mathcal{A}\circ f(x,j) + \sum_{k\neq j}\int_{\mathbf{R}}[f(x+y,k)-f(x,j)]\nu_{jk}B_{jk}\{dy\} \tag{2.121}$$

where

$$\mathcal{A}\circ f(x,j) = d_j\frac{\partial f}{\partial x} + \int_{\mathbf{R}}[f(x+y,k)-f(x,j)-\frac{\partial f}{\partial x}(x,j)\tau(y)]\mu_{jj}\{dy\} \tag{2.122}$$

where d_j *is a constant,* τ *a centering function given by*

$$\tau(x) = \begin{cases} -1 & \text{for } x < -1 \\ |x| & \text{for } |x| \leq 1 \\ 1 & \text{for } x > 1 \end{cases} \tag{2.123}$$

and μ_{jj} *is a Lévy measure.*

Chapter 3

Theory of Semiregenerative Phenomena

In this chapter we develop a theory of semiregenerative phenomena. These may be viewed as a family of linked regenerative phenomena, for which Kingman [54, 55] developed a theory within the framework of quasi-Markov chains. We use a different approach and explore the correspondence between semiregenerative sets and the range of a Markov subordinator with a unit drift, or a Markov renewal process in the discrete-time case.

3.1 Introduction

In the literature there are extensive investigations of semiregenerative processes, which are defined as those having an imbedded Markov renewal process (MRP). Here, the connecting idea is that of the semiregenerative set whose elements are (roughly speaking) the points at which the phenomenon occurs. In discrete time this set coincides with the range of an MRP.

Indeed, the theory of semiregenerative phenomena is a direct extension of Feller's [35] theory of recurrent events (phenomena) in discrete time and Kingman's [55] theory of regenerative phenomena in continuous time. If the observed range of a process contains a semiregenerative set as a subset, then the process is semiregenerative. We take the view that semiregenerative phenomena are important in themselves and therefore worthy of study. For the analysis we use techniques based on results from Markov renewal theory.

[3]This chapter is an updated version of material included in the following two articles: N.U. Prabhu (1988) 'Theory of semiregenerative phenomena', *J. Appl. Probab.* **25A**, pp. 257–274, whose contents are reproduced with permission from the Applied Probability Trust, and N.U. Prabhu (1994), 'Further results for semiregenerative phenomena', *Acta Appl. Math.* **34**, 1-2, pp. 213–223, whose contents are reproduced with permission from Kluwer Academic Publishers.

We present some basic definitions in Sec. 3.2, starting from that of semiregenerative processes, which states that a process $Z = \{Z_{tl}, (t,l) \in T \times E\}$, with $T = \mathbf{R}_+$ or $T = \mathbf{N}_+$ and E being a countable set, is a semiregenerative phenomenon if, in particular, it takes values only the values 0 and 1 and has the following *partial lack of memory* property on the first index with respect to the observation of the value 1:

$$P\{Z_{t_i l_i} = 1\,(1 \leq i \leq r) | Z_{0,l_0} = 1\} = \prod_{i=1}^{r} P\{Z_{t_i - t_{i-1}, l_i} = 1 | Z_{0,l_{i-1}} = 1\}$$

for $0 = t_0 \leq t_1 \leq \cdots \leq t_r$ and $l_0, l_1, \ldots, l_r \in E$. Note that the parameter set of Z, $T \times E$, is the cartesian product of a *time parameter set* T, $\mathbf{R}_+$ or $\mathbf{N}_+$, by a (countable) *label parameter set* E. Thus, the partial lack of memory property may be seen as a temporal one and the observation of the value 1 may be interpreted as the occurrence of a phenomenon of interest.

The discrete time index case $T = \mathbf{N}_+$ leads to a semirecurrent phenomena. In Sec. 3.3 we address semirecurrent phenomena and characterize their relation to MRPs, thus complementing the results on MRPs presented in Chap. 2. The main result obtained is that the semirecurrent set of a semirecurrent phenomena corresponds to the range of an MRP and, conversely, a semirecurrent set can only arise in this manner (Theorem 3.3).

In Sec. 3.4 and Sec. 3.5 we address continuous-time semiregenerative phenomena and related them with Markov-additive processes (MAP), complementing the results for these processes given in Chap. 2. In Sec. 3.4 we construct a Markov subordinator (MAP taking nonnegative values) with a unit drift whose range turns out to be a semiregenerative set (Theorem 3.4). In the case where E is finite we prove the converse in Sec. 3.5, i.e., that every semiregenerative set corresponds to the range of a Markov subordinator (Theorem 3.7).

Our approach yields results analogous to Kingman's ([55], Chapter 5) for quasi-Markov chains. While our approach (based on the property stated above) is thus more rewarding in these respects, our techniques are simpler, being based on properties of MRP. Bondesson [19] has investigated the distribution of occupation times of quasi-Markov processes. We shall not investigate this problem for semiregenerative phenomena.

Professor Erhan Çinlar has remarked to N. U. Prabhu that several papers by him, J. Jacod, H. Kaspi, and B. Maisonneuve on regenerative systems have a bearing on theory presented in this chapter and originally developed in [88]. However, our approach is different from theirs and makes the results more accessible to applied probabilists. In particular, the theory presented here provides a proper perspective to the work of Kulkarni and Prabhu [58] and Prabhu [87], to which no reference is made by the above authors; see Examples 3.3 and 3.5. Moreover, it also lays foundations to the theories of Markov random walks and MAP.

3.2 Basic Definitions

Let the set T be either $\mathbf{R}_+ = [0, \infty)$ or $\mathbf{N}_+ = \{0, 1, 2, \cdots\}$, E denote a countable set and $(\Omega, \mathcal{F}, P)$ a probability space.

Definition 3.1. A semiregenerative phenomenon $Z = \{Z_{tl}, (t, l) \in T \times E\}$ on a probability space $(\Omega, \mathcal{F}, P)$ is a stochastic process taking values 0 or 1 and such that: $\sum_{l \in E} Z_{tl} \leq 1$ for all $t \in T$ a.s., and for all $r \geq 1$ and $(t_i, l_i) \in T \times E \, (1 \leq i \leq r)$, with $0 = t_0 \leq t_1 \leq \cdots \leq t_r$, and $j \in E$, we have

$$
\begin{aligned}
P\{Z_{t_1 l_1} =& Z_{t_2 l_2} = \cdots = Z_{t_r l_r} = 1 | Z_{0j} = 1\} \\
&= \prod_{i=1}^{r} P\{Z_{t_i - t_{i-1}, l_i} = 1 | Z_{0, l_{i-1}} = 1\} \quad (l_0 = j). \qquad (3.1)
\end{aligned}
$$

Remark 3.1. The previous definition includes the condition $\sum_{l \in E} Z_{tl} \leq 1$, which is introduced to avoid the possibility that, for a fixed value of t, $Z_{tl} = 1$ for more than one value of l. This turns 'special' the observation of the value 1 for the random variable Z_{tl}, implying in particular in such case that $Z_{tl'} = 0$ for all $l' \neq l$. The condition $\sum_{l \in E} Z_{tl} \leq 1$ may fail with all other conditions of the definition holding. To see this fact, consider a process $Z = \{Z_{tl}, (t, l) \in T \times E\}$, where $E = \{1, 2, \ldots, N\}$ with $N > 1$, such that

$$P\{Z_{tl} = 1, \forall (t, l) \in T \times E\} = 1/2 = P\{Z_{tl} = 0, \forall (t, l) \in T \times E\}.$$

It is easy to see that this process satisfies (3.1) since

$$P\{Z_{tl} = 1 | Z_{0j} = 1\} = 1, \quad \text{for all } j, l \in E \text{ and } t \geq 0$$

but, however, $P\{\sum_{l \in E} Z_{tl} = N\} = 1/2$, for all $t \in T$.

For each $l \in E$, denote $Z_l = \{Z_{tl}, t \in T\}$. Since

$$
\begin{aligned}
P\{Z_{t_1 l} &= Z_{t_2 l} = \cdots = Z_{t_r l} = 1 | Z_{0j} = 1\} \\
&= P\{Z_{t_1 l} = 1 | Z_{0j} = 1\} \prod_{i=2}^{r} P\{Z_{t_i - t_{i-1}, l} = 1 | Z_{0l} = 1\}, \qquad (3.2)
\end{aligned}
$$

Z_l is a (possibly delayed) regenerative phenomenon in the sense of Kingman [55] in the continuous-time case $T = \mathbf{R}_+$, and a recurrent event (phenomenon) in the sense of Feller [35] in the discrete time case $T = \mathbf{N}_+$.

The family $Z' = \{Z_l, l \in E\}$ is a family of linked regenerative phenomena, for which a theory was developed by Kingman [54] in the case of finite E; later he reformulated the results in terms of quasi-Markov chains [55]. This concept is explained below.

Example 3.1. Let $J = \{J_t, t \in T\}$ be a time-homogeneous Markov chain on the state space E and denote

$$Z_{tl} = \mathbf{1}_{\{J_t = l\}} \quad \text{for } (t, l) \in T \times E. \tag{3.3}$$

The random variables Z_{tl} satisfy the relation (3.1), which is merely the Markov property. More generally, let C be a fixed subset of E and

$$Z_{tl} = \mathbf{1}_{\{J_t = l\}} \quad \text{for } (t, l) \in T \times C. \tag{3.4}$$

These random variables also satisfy (3.1) and thus $Z = \{Z_{tl},\ (t, l) \in T \times C\}$ is a semiregenerative phenomenon. In particular, suppose that C is a finite subset of E and define

$$K_t = \begin{cases} J_t & \text{if } J_t \in C \\ 0 & \text{if } J_t \notin C. \end{cases} \tag{3.5}$$

Then $\{K_t,\ t \in T\}$ is defined to be a quasi-Markov chain on the state space $C \cup \{0\}$.

While the quasi-Markov chain does provide a good example of a semiregenerative phenomenon (especially in the case of finite E), it does not reveal the full features of these phenomena; in particular, it does not establish their connection with Markov additive processes.

Definition 3.2. We let

$$\zeta = \{(t, l) \in T \times E : \ Z_{tl} = 1\} \tag{3.6}$$

and call ζ the semiregenerative set associated with Z.

The main theme of this chapter is the correspondence between the set ζ and the range of a Markov renewal process (in the discrete-time case) and of a Markov subordinator with a unit drift (in the continuous-time case). Kingman ([55], p. 123) has remarked that associated with a quasi-Markov chain there is a process of type F studied by Neveu [74]. The Markov subordinator we construct for our purpose is indeed a process of type F, but we concentrate on properties of the range of this process. For a detailed description of Markov-additive processes see Çinlar [28, 29].

To complete Definition 3.1, we specify the initial distribution $\{a_j, j \in E\}$, where

$$P\{Z_{0j} = 1\} = a_j \tag{3.7}$$

with $a_j \geq 0$ and $\sum a_j = 1$. As in the case of regenerative phenomena, it can be proved that the relation (3.1) determines all finite-dimensional distributions of Z and that Z is strongly regenerative (that is, (3.1) holds for stopping times). We shall write P_j and E_j for the probability and the expectation conditional on the event $\{Z_{0j} = 1\}$.

In the discrete-time case we call Z a semirecurrent phenomenon and denote

$$u_{jk}(n) = P\{Z_{nk} = 1 | Z_{0j} = 1\} \tag{3.8}$$

where $u_{jk}(0) = \delta_{jk}$. In the continuous-time case we let

$$P_{jk}(t) = P\{Z_{tk} = 1 | Z_{0j} = 1\} \tag{3.9}$$

where $P_{jk}(0) = \delta_{jk}$. The phenomenon is standard if

$$P_{jk}(t) \to \delta_{jk} \quad \text{as} \quad t \to 0^+. \tag{3.10}$$

In this case it is known that the limit

$$\lim_{t \to 0^+} \frac{1 - P_{jj}(t)}{t} \quad (j \in E) \tag{3.11}$$

exists but may be possibly infinite. We shall only consider stable semiregenerative phenomena, where the limit (3.11) is finite for every $j \in E$.

3.3 Semirecurrent Phenomena

In this section we address semirecurrent phenomena and characterize their relation to Markov renewal processes. The main result of the section is that the semirecurrent set of a semirecurrent phenomena corresponds to the range of an MRP and, conversely, a semirecurrent set can only arise in this manner (Theorem 3.3). For the results from Markov renewal theory used in the chapter see Chap. 2 and Çinlar ([30], Chapter 10).

We let $\mathcal{L} = \mathbf{N}_+ \times E$ and ζ be the semirecurrent set defined by (3.6).

Definition 3.3. Let $T_0 = 0$ and for $r \geq 1$

$$T_r = \min\{n > T_{r-1} : (n, l) \in \zeta \text{ for some } l\}. \tag{3.12}$$

We shall call T_r the semirecurrence times of Z.

Let $J_r = l$ when $Z_{T_r,l} = 1$. Definition 3.1 shows that this l is unique. We have the following.

Theorem 3.1. *The process $\{(T_r, J_r),\ r \geq 0\}$ is an MRP on the state space $\mathcal{L}$.*

Proof. The desired result follows from the fact that

$$\begin{aligned} P\{T_{r+1} = n, J_{r+1} = l | J_0, T_1, J_1, \cdots, T_r, J_r\} \\ = P_{J_r}\{T_{r+1} - T_r = n - T_r, J_{r+1} = l\} \quad \text{a.s.} \end{aligned} \tag{3.13}$$

□

Following the previous result, we shall denote by P_j the conditional probability given $J_0 = j$.

Remark 3.2. Let

$$\mathcal{R} = \{(n,l) \in \mathbf{N}_+ \times E : (T_r, J_r) = (n,l) \text{ for some } r \geq 0\}$$

be the observed range of the process $\{T_r, J_r\}$ and, as in (2.27),

$$N = \#\{r > 0 : (T_r, J_r) \in \mathcal{R}\}$$

denote the maximum number of observed transitions. Since $T_{r+1} > T_r$ a.s., $T_r \to \infty$ as $r \to \infty$ and so

$$[N = \infty] \Longrightarrow \left[\sigma = \sup_{0 \leq r \leq N} T_r = \infty\ a.s.\right].$$

Therefore the case $N = \infty$ and $\sigma < \infty$, which corresponds to case (c) of Sec. 2.5, does not arise in the MRP $\{(T_r, J_r)\}$.

According to the previous remark, we propose the following characterization of Z.

Definition 3.4. We say that the semirecurrent phenomenon Z (or the associated set J) is terminating or nonterminating according as its maximum number of observed transitions is finite ($N < \infty$) or infinite ($N = \infty$).

We shall denote by

$$q_{jk}^{(r)}(n) = P_j\{T_r = n, J_r = k\} \tag{3.14}$$

the semirecurrence time distribution of Z. We have $q_{jk}^{(0)}(0) = \delta_{jk}$ and $q_{jk}^{(0)}(n) = 0$ for $n \geq 1$. We shall write $q_{jk}^{(1)}(n) = q_{jk}(n)$. On account of the Markov renewal property, we have for $r, s \geq 0$

$$q_{jk}^{(r+s)}(n) = \sum_{m=0}^{n} \sum_{l \in E} q_{jl}^{(r)}(m)\, q_{lk}^{(s)}(n-m). \tag{3.15}$$

Theorem 3.2. *The probabilities* $\{u_{jk}(n), n \in \mathbf{N}_+, j, k \in E\}$ *defined by (3.8) form the unique solution of the equations*

$$x_{jk}(n) = q_{jk}(n) + \sum_{m=1}^{n-1} \sum_{l \in E} q_{jl}(m) x_{lk}(n-m) \tag{3.16}$$

with $0 \leq x_{jk}(n) \leq 1$. *This solution is given by* $u_{jk}(0) = q_{jk}(0) = \delta_{jk}$ *and*

$$u_{jk}(n) = \sum_{r=1}^{n} q_{jk}^{(r)}(n) \quad (n \geq 1). \tag{3.17}$$

Proof. As the result follows trivially for $n = 0$, we assume that $n \geq 1$. From (3.6) and (3.12) it follows that for $n \geq 1$,

$$\begin{aligned} u_{jk}(n) &= P_j\{(T_r, J_r) = (n, k) \text{ for some } r \geq 1\} \\ &= \sum_{r=1}^{n} q_{jk}^{(r)}(n), \end{aligned} \tag{3.18}$$

the sum going only up to $r = n$ since $T_r \geq r$ a.s. Thus the $u_{jk}(n)$ are given by (3.17). In addition, we have

$$\begin{aligned} u_{jk}(n) &= q_{jk}^{(1)}(n) + \sum_{t=1}^{n-1} q_{jk}^{(t+1)}(n) \\ &= q_{jk}(n) + \sum_{t=1}^{n-1} \sum_{m=1}^{n-1} \sum_{l \in E} q_{jl}(m) q_{lk}^{(t)}(n-m) \end{aligned}$$

using (3.15). Thus

$$u_{jk}(n) = q_{jk}(n) + \sum_{m=1}^{n-1} \sum_{l \in E} q_{jl}(m) u_{lk}(n-m) \tag{3.19}$$

in view of (3.18). To prove the uniqueness of the solution of (3.16) we find that if $\{x_{jk}(n)\}$ is a solution of (3.16), then

$$\begin{aligned} x_{jk}(n) &= q_{jk}^{(1)}(n) + \sum_{m=1}^{n-1} \sum_{l \in E} q_{jl}(m) \Bigg[q_{lk}(n-m) \\ &\quad + \sum_{m'=1}^{n-m-1} \sum_{l' \in E} q_{ll'}(n-m-m') x_{l'k}(m') \Bigg] \\ &= q_{jk}^{(1)}(n) + q_{jk}^{(2)}(n) + \sum_{m'=1}^{n-1} \sum_{l' \in E} q_{jl'}^{(2)}(n-m') x_{l'k}(m'). \end{aligned}$$

By induction we find that

$$x_{jk}(n) = \sum_{t=1}^{r} q_{jk}^{(t)}(n) + \sum_{m=1}^{n-1} \sum_{l \in E} q_{jl}^{(r)}(m) x_{lk}(n-m).$$

Since $q_{jk}^{(r)}(n) = 0$ for $r > n$ we have

$$x_{jk}(n) = \sum_{t=1}^{n} q_{jk}^{(t)}(n) = u_{jk}(n),$$

which proves the uniqueness of the solution (3.17). □

With respect to the theory of MRPs, we should note that the previous theorem is a consequence of theorems 2.3 and 2.6, in view of the previous conclusion in Remark 3.2 that case (c) of Sec. 2.5 does not arise in the MRP $\{(T_r, J_r)\}$.

The previous theorem establishes that a semirecurrent phenomenon gives rise to an MRP with the sojourn times $T_r - T_{r-1}$ $(r \geq 1)$ concentrated on the set $\{1, 2, \cdots\}$. The following result shows that this is the only way that a semirecurrent phenomenon can occur.

Theorem 3.3. (a). *Let $\{(T_r, J_r), r \geq 0\}$ be an MRP on the state space $\mathcal{L}$, with the sojourn time distribution concentrated on $\{1, 2, \ldots\}$,*

$$\mathcal{R}' = \{(n, l) \in \mathcal{L} : (T_r, J_r) = (n, l) \text{ for some } r \geq 0\} \tag{3.20}$$

denote the range of the MRP, and $Z'_{nl} = \mathbf{1}_{\{(n,l) \in \mathcal{R}'\}}$.

Then the process $Z' = \{Z'_{nl}, (n, l) \in \mathcal{L}\}$ is a semirecurrent phenomenon. (b). *Conversely, any semirecurrent phenomenon Z is equivalent to a phenomenon Z' generated in the above manner in the sense that Z and Z' have the same $\{u_{jk}(n)\}$ sequence.*

Proof. (a). We have $P\{Z'_{0j} = 1\} = P\{J_0 = j\}$. Let $v_{jk}(0) = \delta_{jk}$ and

$$v_{jk}(n) = P_j\{Z'_{nk} = 1\} \quad (n \geq 1).$$

Then, for $n \geq 1$,

$$v_{jk}(n) = P_j\{(n, k) \in \mathcal{R}\} = \sum_{r=1}^{n} P_j\{T_r = n, J_r = k\}. \tag{3.21}$$

Let $r \geq 1$ and denote

$$A_r = \{\mathbf{t} = (t_1, t_2, \ldots, t_r) : 0 \leq t_1 \leq t_2 \leq \ldots, \leq t_r\}.$$

Then, for $0 = n_0 \leq n_1 \leq n_2 \leq \cdots \leq n_r$, $l_1, l_2, \ldots, l_r, j \in E$, we have, with $l_0 = j$ and $s_0 = t_0 = 0$,

$$\begin{aligned}
P\{Z'_{n_1 l_1} &= Z'_{n_2 l_2} = \cdots = Z'_{n_r l_r} = 1 | Z'_{0j} = 1\} \\
&= P_j \left\{ \bigcap_{i=1}^{r} \bigcup_{s_i \geq s_{i-1}} \{(T_{s_i}, J_{s_i}) = (n_i, l_i)\} \right\} \\
&= P_j \left\{ \bigcup_{\mathbf{t} \in A_r} \bigcap_{i=1}^{r} \{(T_{t_i}, J_{t_i}) = (n_i, l_i)\} \right\} \\
&= \sum_{\mathbf{t} \in A_r} \prod_{i=1}^{r} P\{(T_{t_i}, J_{t_i}) = (n_i, l_i) | (T_{t_{i-1}}, J_{t_{i-1}}) = (n_{i-1}, l_{i-1})\} \\
&= \sum_{\mathbf{t} \in A_r} \prod_{i=1}^{r} P_{l_{i-1}} \{(T_{t_i - t_{i-1}}, J_{t_i - t_{i-1}}) = (n_i - n_{i-1}, l_i)\} \\
&= \prod_{i=1}^{r} \sum_{t_i \geq t_{i-1}} P_{l_{i-1}} \{(T_{t_i - t_{i-1}}, J_{t_i - t_{i-1}}) = (n_i - n_{i-1}, l_i)\} \\
&= \prod_{i=1}^{r} v_{l_{i-1} l_i}(n_i - n_{i-1}) \\
&= v_{j l_1}(n_1) v_{l_1 l_2}(n_2 - n_1) \cdots v_{l_{r-1} l_r}(n_r - n_{r-1}).
\end{aligned}$$

This shows that Z' is a semirecurrent phenomenon. Note that, from the Markov renewal property,

$$\begin{aligned}
q_{jk}^{(r+1)}(n) &= P_j\{T_{r+1} = n, J_{r+1} = k\} \\
&= \sum_{m=1}^{n-1} \sum_{l \in E} P_j\{T_1 = m, J_1 = l\} \cdot P_l\{T_r = n - m, J_r = k\} \\
&= \sum_{m=1}^{n-1} \sum_{l \in E} q_{jl}(n)\, q_{lk}^{(r)}(n),
\end{aligned}$$

which, in view of (3.21), leads to the relation

$$v_{jk}(n) = q_{jk}(n) + \sum_{m=1}^{n-1} \sum_{l \in E} q_{jl}(m) v_{lk}(n - m) \quad (n \geq 1) \tag{3.22}$$

where $q_{jk}(n) = P_j\{T_1 = n, J_1 = k\}$.

(b). Conversely, let Z be a semirecurrent phenomenon with the associated sequence $\{u_{jk}(n), (n, k) \in \mathcal{L}\}$. Let $\{q_{jk}(n)\}$ be the associated semirecurrence time distribution, so that by Theorem 3.2, Equation (3.19) holds.

Let Z' be the semirecurrent phenomenon constructed as in (a) from the sojourn time distribution $\{q_{jk}(n)\}$. Then $\{v_{jk}(n)\}$ satisfies (3.22). Because of the uniqueness of the solution of (3.16) we find that $\{v_{jk}(n) = u_{jk}(n)\}$, as required. □

Example 3.1 (continuation). For a quasi-Markov chain in discrete time the semirecurrence times are the hitting times of the set C. If $K_0 = j \in C$, the process spends one unit of time in j and $T_1 - 1$ units outside of C before returning to C. If $C = E$, then $T_r = r$ almost surely (a.s.) for all $r \geq 0$.

In the rest of the section we provide some illustrative examples of semirecurrent phenomena.

Example 3.2. Let $\{(K_n, J_n), n \in \mathbf{N}_+\}$ be a time-homogeneous Markov chain on the state space $S \times E$ (with S arbitrary). Let $a \in S$ (fixed) and assume that $K_0 = a$ a.s. Define

$$Z_{nl} = \mathbf{1}_{\{K_n = a, J_n = l\}} \qquad ((n, l) \in \mathcal{L}).$$

On account of the Markov property, $Z = \{Z_{nl}\}$ is a semirecurrent phenomenon. The semirecurrence times are the successive hitting times of the line $K_n = a$.

Example 3.3. Let $\{X_n, n \in \mathbf{N}_+\}$ be a Markov chain on the state space $\mathbf{N}_+$, and

$$M_n = \max(X_0, X_1, X_2, \cdots, X_n) \qquad (n \in \mathbf{N}_+)$$

be its maximum function. Also define

$$Z_{nl} = \mathbf{1}_{\{X_n = M_n = l\}} \qquad ((n, l) \in \mathcal{L}).$$

We have $P\{Z_{0j} = 1\} = P\{X_0 = j\}$, and, for $r \geq 1$, $0 = n_0 \leq n_1 \leq n_2 \leq \cdots \leq n_r$, $0 \leq j = l_0 \leq l_1 \leq l_2 \leq \ldots \leq l_r$, using the Markov property

$$\begin{aligned}
&P\{Z_{n_1 l_1} = Z_{n_2 l_2} = \cdots = Z_{n_r l_r} = 1 | Z_{0j} = 1\} \\
&= P_j \left\{ \bigcap_{i=1}^{r} \{X_{n_{i-1}+t} \leq X_{n_i} (0 \leq t \leq n_i - n_{i-1}), X_{n_i} = l_i\} \right\} \\
&= \prod_{i=1}^{r} P\{X_{n_{i-1}+t} \leq X_{n_i} (0 \leq t \leq n_i - n_{i-1}), X_{n_i} = l_i | X_{n_{i-1}} = l_{i-1}\} \\
&= \prod_{i=1}^{r} P\{X_t \leq X_{n_i - n_{i-1}} (0 \leq t \leq n_i - n_{i-1}), X_{n_i - n_{i-1}} = l_i | X_0 = l_{i-1}\} \\
&= \prod_{i=1}^{r} P\{Z_{n_i - n_{i-1}, l_i} = 1 | Z_{0 l_{i-1}} = 1\}.
\end{aligned}$$

This shows that $Z = \{Z_{nl}\}$ is a semirecurrent phenomenon with

$$\begin{aligned} u_{jk}(n) &= P\{Z_{nk} = 1 | Z_{0j} = 1\} \\ &= P\{X_t \leq X_n (0 \leq t \leq n), X_n = k | X_0 = j\}. \end{aligned}$$

The semirecurrence times of this phenomenon are called ascending ladder epochs by Kulkarni and Prabhu [58], who study the fluctuation theory of the Markov chain.

Example 3.4. Let$\{(S_n, J_n), n \in \mathbf{N}_+\}$ be a Markov random walk (discrete-time version of Markov-additive process) on the state space $\mathbf{R} \times E$ and

$$M_n = \max(0, S_1, S_2, \ldots, S_n) \quad (n \in \mathbf{N}_+)$$

its maximum functional. Denote

$$Z_{nl} = \mathbf{1}_{\{S_n = M_n, J_n = l\}} \quad ((n, l) \in \mathcal{L}).$$

We have $P\{Z_{0j} = 1\} = P\{J_0 = j\}$, and, for $r \geq 1$, $0 = n_0 \leq n_1 \leq n_2 \leq \cdots \leq n_r$, $j, l_1, l_2, \ldots, l_r \in E$, with $l_0 = j$,

$$\begin{aligned} &P\{Z_{n_1 l_1} = Z_{n_2 l_2} = \cdots = Z_{n_r l_r} = 1 | Z_{0j} = 1\} \\ &= P_j \left\{ \bigcap_{i=1}^{r} \{S_{n_{i-1}+t} \leq S_{n_i} (0 \leq t \leq n_i - n_{i-1}), J_{n_i} = l_i\} \right\} \\ &= P_j \left\{ \bigcap_{i=1}^{r} \{S_{n_{i-1}+t} - S_{n_{i-1}} \leq S_{n_i} - S_{n_{i-1}} (0 \leq t \leq n_i - n_{i-1}), J_{n_i} = l_i\} \right\} \\ &= \prod_{i=1}^{r} P\{S_t \leq S_{n_i - n_{i-1}} (0 \leq t \leq n_i - n_{i-1}), J_{n_i - n_{i-1}} = l_i | J_0 = l_{i-1}\} \\ &= \prod_{i=1}^{r} P\{Z_{n_i - n_{i-1}, l_i} = 1 | Z_{0, l_{i-1}} = 1\}. \end{aligned}$$

This shows that $Z = \{Z_{nl}\}$ is a semirecurrent phenomenon with

$$\begin{aligned} u_{jk}(n) &= P\{Z_{nk} = 1 | Z_{0j} = 1\} \\ &= P\{S_t \leq S_n (0 \leq t \leq n), J_n = k | J_0 = j\}. \end{aligned}$$

We obtain a second such phenomenon by using the minimum functional, $m_n = \min(0, S_1, S_2, \ldots, S_n)$, $n \in \mathbf{N}_+$, and considering

$$Z'_{nl} = \mathbf{1}_{\{S_n = m_n, J_n = l\}} \quad ((n, l) \in \mathcal{L}).$$

These two phenomena determine the fluctuation behavior of the random walk.

3.4 A Continuous-Time Semiregenerative Phenomenon

In this and the next section we address continuous-time semiregenerative phenomena. The main contribution of this section is the construction a Markov subordinator with a unit drift whose range turns out to be a semiregenerative set (Theorem 3.4).

We let $\mathcal{L} = \mathbf{R}_+ \times E$ and $J = \{J(\tau),\, \tau \geq 0\}$ be a time-homogeneous Markov process on the state space E, all of whose states are stable. Let $T_0 = 0$ and $T_n\,(n \geq 1)$ be the epochs of successive jumps in J; for convenience we denote $J_n = J(T_n)\,(n \geq 0)$. We define a sequence of continuous-time processes $\{X_n^{(1)},\, n \geq 1\}$ and a sequence of random variables $\{X_n^{(2)},\, n \geq 1\}$ as follows.

(i) On $\{J_n = j\}$, $\{X_{n+1}^{(1)}(\tau),\, 0 \leq \tau \leq T_{n+1} - T_n\}$ is a subordinator with a unit drift and Lévy measure $\mu_{jj}^{(0)}$.

(ii) Given J_0 and $\{(X_m^{(1)}, X_m^{(2)}, J_m), 1 \leq m \leq n\}$, the increment process $\{X_{n+1}^{(1)}(\tau - T_n), T_n \leq \tau \leq T_{n+1}\}$ and the pair of random variables $(X_{n+1}^{(2)}, J_{n+1})$ depend only on J_n.

(iii) Given J_n, $\{X_{n+1}^{(1)}(\tau),\, 0 \leq \tau \leq T_{n+1} - T_n\}$ and $(X_{n+1}^{(2)}, J_{n+1})$ are conditionally independent, with respective distribution measures

$$H_j\{\tau; A\} \quad \text{and} \quad \lambda_{jk} F_{jk}\{A\} \quad (j \neq k) \tag{3.23}$$

for any Borel subset A of $\mathbf{R}_+$. Here the F_{jk} are concentrated on $\mathbf{R}_+$, while H_j is concentrated on $[0, \infty]$; $\lambda_{jk}\,(j \neq k)$ are the transition rates of the process J and we denote $\lambda_{jj} = \sum_{k \neq j} \lambda_{jk}\,(0 < \lambda_{jj} < \infty)$.

According to this construction, on intervals where $J(r) = j$, the process X evolves as an ordinary subordinator with Lévy measure $\mu_{jj}^{(0)}$ depending on j. In addition, when J jumps from j to k, X receives a jump with distribution F_{jk}; these are the Markov-modulated jumps.

Let us now address the process $\{(S_n, J_n),\, n \geq 0\}$ with

$$S_0 = 0, \quad S_n = \sum_{i=1}^{n} [X_i^{(1)}(T_i - T_{i-1}) + X_i^{(2)}] \quad (n \geq 1). \tag{3.24}$$

From the above conditions it follows that $\{(S_n, J_n)\}$ is an MRP on the state space $\mathbf{R}_+ \times E$, whose transition distribution measure

$$P\{S_{n+1} \in A, J_{n+1} = k | S_n = s, J_n = j\} = Q_{jk}\{A - s\}$$

is given by

$$Q_{jk}\{A\} = \int_0^\infty \int_{0-}^{\infty^+} \exp(-\lambda_{jj} s) \lambda_{jk} H_j\{s; dx\} F_{jk}\{A - x\}\, ds \tag{3.25}$$

for $k \neq j$, and $Q_{jj}\{A\} = 0$. We denote by $U_{jk}\{A\}$ the Markov renewal measure associated with this process, so that

$$U_{jk}\{A\} = \sum_{n=0}^{\infty} P_j\{S_n \in A, J_n = k\} \tag{3.26}$$

with P_j denoting the conditional probability given $J_0 = j$ (or $J(0) = j$ for that matter).

We construct a process $(Y, J) = \{(Y(\tau), J(\tau)), \tau \geq 0\}$ as follows:

$$(Y(\tau), J(\tau)) = \begin{cases} (S_n + X_{n+1}^{(1)}(\tau - T_n), J_n) & \text{for } T_n \leq \tau < T_{n+1} \\ (L, \Delta) & \text{for } \tau \geq L' \end{cases} \tag{3.27}$$

where

$$L' = \sup_{n\geq 0} T_n \quad \text{and} \quad L = \sup_{n\geq 0} S_n \tag{3.28}$$

(so that, in particular, $L' \leq L \leq \infty$) and Δ is a point of compactification of the set E. Denoting

$$N(\tau) = \max\{n : T_n \leq \tau\}$$

we can write

$$(Y(\tau), J(\tau)) = (S_{N(\tau)} + X_{N(\tau)+1}^{(1)}(\tau - T_{N(\tau)}), J_{N(\tau)}) \quad (\tau < L') \tag{3.29}$$

and in view of the assumptions (i)-(iii) above we find that

$$\begin{aligned} P\{Y(\tau + \tau') \in A, J(\tau + \tau') = k|(Y(s), J(s)) = (y_s, j_s), 0 \leq s \leq \tau\} \\ = P\{S_{N(\tau+\tau')} - S_{N(\tau)} + X_{N(\tau+\tau')+1}^{(1)}(\tau + \tau' - T_{N(\tau+\tau')}) \\ - X_{N(\tau)+1}^{(1)}(\tau - T_{N(\tau)}) \in A - Y(\tau), J(\tau + \tau') = k \,| \\ (Y(\tau), J(\tau)) = (y_\tau, j_\tau)\} \\ = P\{S_{N(\tau')} + X_{N(\tau')+1}^{(1)}(\tau' - T_{N(\tau')}) \in A - y_\tau, J(\tau') = k|J(0) = j_\tau\} \\ = P\{Y(\tau') \in A - y_\tau, J(\tau') = k|J(0) = j_\tau\}. \end{aligned}$$

This shows that (Y, J) is an MAP on the state space $\mathbf{R}_+ \times E$.

Let $f(t, j)$ be a bounded function on $\mathbf{R}_+ \times E$ such that for each fixed j, f is continuous and has a bounded continuous derivative $\partial f/\partial t$. Then the infinitesimal generator $\mathcal{A}$ of (Y, J) is found to be

$$\begin{aligned} (\mathcal{A}f)(t, j) = \frac{\partial f}{\partial t} + \int_{0-}^{\infty^+} [f(t + v, j)) - f(t, j)]\mu_{jj}^{(0)}\{dv\} \\ + \sum_{k\neq j} \int_{0-}^{\infty} [f(t + v, k) - f(t, j)]\lambda_{jk}F_{jk}\{dv\}. \end{aligned} \tag{3.30}$$

This shows that the jumps in Y are those in the additive component plus the Markov-modulated jumps with the distribution measures $\lambda_{jk}F_{jk}(k \neq j)$.

The transition distribution measure of (Y, J) can be expressed in terms of the Markov renewal measure associated with the MRP $\{(S_n, T_n, J_n),\ n \geq 0\}$. We are interested in the range of the process (Y, J),

$$\mathcal{R} = \{(t, l) \in \mathbf{R}_+ \times E : (Y(\tau), J(\tau)) = (t, l) \text{ for some } \tau \geq 0\}. \tag{3.31}$$

We define a process $Z = \{Z_{tl},\ (t, l) \in \mathcal{L}\}$ by

$$Z_{tl} = \mathbf{1}_{\{(t,l)\in\mathcal{R}\}} \quad ((t, l) \in \mathcal{L}) \tag{3.32}$$

and, following (3.9), let $P_{jk}(t) = P\{Z_{tk} = 1|Z_{0j} = 1\}$. An inspection of the sample path of Z shows that we can write the range $\mathcal{R}$ as

$$\mathcal{R} = \bigcup_{n=0}^{\infty} \left[\left(S_n + \mathcal{R}_{X_{n+1}^{(1)}}\right) \times \{J_n\}\right], \tag{3.33}$$

where $\mathcal{R}_{X_{n+1}^{(1)}}$ is the range of the subordinator $X_{n+1}^{(1)}$ and, as usual, for sets A and B on a vector space $A+B = \{a+b : a \in A, b \in B\}$. We know that each $\mathcal{R}_{X_{n+1}^{(1)}}$ is a regenerative set (Kingman [55], Section 4.2) with p-function

$$p_j(t) = P\left\{t \in \mathcal{R}_{X_{n+1}^{(1)}} | J_n = j\right\} \tag{3.34}$$

satisfying

$$\begin{aligned} \int_A p_j(t)dt &= E\left[\int_0^{T_1} H_j\{s; A\}ds\right] \\ &= \int_0^{\infty} \exp(-\lambda_{jj}s)H_j\{s; A\}\, ds \end{aligned} \tag{3.35}$$

for any Borel subset A of $\mathbf{R}_+$. We have then the following.

Theorem 3.4. *The family $Z = \{Z_{tl}, (t, l) \in \mathbf{R}_+ \times E\}$ is a standard semiregenerative phenomenon, for which*

$$P_{jk}(t) \geq \int_{0-}^{t} U_{jk}\{ds\}p_k(t - s). \tag{3.36}$$

The equality in (3.36) holds iff $L = \infty$ a.s.

Proof. We have $P\{Z_{0j} = 1\} = P\{J_0 = j\}$. The semiregenerative property (3.1) can be established exactly as in the proof of Theorem 3.3(a) using the Markov-additive property of (Y, J). We have

$$P_{jk}(t) = P_j\left\{(t,k) \in \mathcal{R}_{X_1^{(1)}} \times \{J_0\}\right\} +$$

$$\sum_{l\in E}\int_{0-}^{t} P_j\{S_1 \in ds, J_1 = l\} \times P\{(t,k) \in \mathcal{R}|S_1 = s, J_1 = l\}$$

$$= p_j(t)\delta_{jk} + \sum_{l\in E}\int_{0-}^{t} Q_{jl}\{ds\}P\{(t-s,k) \in \mathcal{R}|J_0 = l\}$$

$$= p_j(t)\delta_{jk} + \sum_{l\in E}\int_{0-}^{t} Q_{jl}\{ds\}P_{lk}(t-s).$$

From the previous statement, $P_{jk}(t)$ satisfies the integral equation

$$G_{jk}(t) = h_{jk}(t) + \sum_{l\in E}\int_{0-}^{t} Q_{jl}\{ds\}G_{lk}(t-s), \tag{3.37}$$

with $h_{jk}(t) = p_j(t)\,\delta_{jk}$. We seek a solution $G_{jk}(t)$ such that for fixed $j, k \in E$, G_{jk} is bounded over finite intervals, and for each $t \in \mathbf{R}_+$, G_{jk} is bounded. The inequality (3.36) follows from the fact that the minimal solution of (3.37) is given by

$$\sum_{l\in E}\int_{0-}^{t} U_{jl}\{ds\}h_{lk}(t-s) = \int_{0-}^{t} U_{jk}\{ds\}p_k(t-s).$$

The solution is unique iff $L = \infty$ (see Theorem 2.6 along with Theorem 2.3). From (3.36) we find, in particular, that

$$P_{jj}(t) \geq p_j(t) \to 1 \quad \text{as} \quad t \to 0^+ \quad (j \in E). \tag{3.38}$$

This shows that Z is a standard phenomenon. □

From the definition of (Y, J) process it is clear that the lifetime of the phenomenon Z is given by

$$\sup\{t : Z_{tl} = 1 \text{ for some } l \in E\} = L \leq \infty \quad a.s. \tag{3.39}$$

and the total duration (occupation time) of Z by

$$\sum_{l\in E}\int_{0}^{L} Z_{tl}dt \geq L' \quad a.s. \tag{3.40}$$

The following results follow from (3.36) in the case $L = \infty$. We have then

$$P_{jk}(t) = \int_{0-}^{t} U_{jk}\{ds\}p_k(t-s). \tag{3.41}$$

We first note the following. Let

$$N = \min\{n \geq 1 : J_n = k\} \quad \text{and} \quad G_{jk}\{A\} = P_j\{S_N \in A\}. \tag{3.42}$$

Using the relation

$$U_{jk}\{A\} = \int_0^\infty G_{jk}\{ds\} U_{kk}\{A - s\} \quad (j \neq k) \tag{3.43}$$

in (3.41), we can write

$$P_{jk}(t) = \int_0^t G_{jk}\{ds\} P_{kk}(t - s) \quad (j \neq k); \tag{3.44}$$

(cf. Kingman [55], Theorem 5.3). The regularity properties of $P_{jk}(t)$ and its asymptotic behavior follow from (3.41)-(3.44).

Theorem 3.4 shows that the range of the MAP constructed above is a semiregenerative set. The converse statement is that every semiregenerative set corresponds to the range of a Markov subordinator with unit drift. We are able to prove this only in the case of the Markov chain J having a finite state space (Theorem 3.7 below). However, the following examples show that the converse is true in two important cases.

Example (3.1) (continuation). Let $\{J_t\}$ be the Markov chain of this example, with $T = \mathbf{R}_+$. Assume that all of its states are stable and use the notation of this section. Let Z_{tl} be as defined by (3.3). We have already observed that $Z = \{Z_{tl}\}$ is a semiregenerative phenomenon. An inspection of the sample path of J shows that the semiregenerative set of Z is

$$\zeta = \bigcup_{n=0}^{\infty} \{[T_n, T_{n+1}) \times \{J_n\}\}. \tag{3.45}$$

This set is the range of the Markov subordinator $(Y, J) = \{\tau, J(\tau)\}$. This means that $X_{n+1}^{(1)}(\tau - T_n) = \tau - T_n$ $(T_n \leq \tau \leq T_{n+1})$ and $F_{jk}\{0\} = 1$ $(k \neq j)$, so that the transition distribution measure Q_{jk} has density

$$\exp(-\lambda_{jj}s)\lambda_{jk} \quad (k \neq j). \tag{3.46}$$

Since $\mathcal{R}_{X_1^{(1)}} = [0, T_1)$ we find that

$$p_j(t) = P\{T_1 > t | J_0 = j\} = \exp(-\lambda_{jj}t) \tag{3.47}$$

and

$$P_{jk}(t) \geq \int_{0-}^t U_{jk}\{ds\} \exp(-\lambda_{kk}(t - s)). \tag{3.48}$$

The equality in (3.48) holds iff $L' = \infty$ a.s.

Example 3.5. This is the continuous-time version of Example 3.3. Let $X = \{X(t),\, t \geq 0\}$ be a continuous Markov chain on $\mathbf{N}_+$, all of whose states are stable. Let now define the process $(Y, J) = \{(Y(\tau), J(\tau)),\, \tau \geq 0\}$ by

$$Y(\tau) = \inf\{t \geq 0 : L(t) > \tau\} \quad \text{and} \quad J(\tau) = M(Y(\tau)) \tag{3.49}$$

where

$$L(t) = \text{ Lebesgue measure of } \{s \geq 0 : X(s) = M(s)\} \cap (0, t] \tag{3.50}$$

and

$$M(t) = \sup_{0 \leq s \leq t} X(s). \tag{3.51}$$

Prabhu [87] showed that the process (Y, J) is a Markov subordinator on the state space $\mathbf{R}_+ \times \mathbf{N}_+$, with the infinitesimal generator $\mathcal{A}$ given by

$$\begin{aligned}(\mathcal{A}f)(t,j) = \frac{\partial f}{\partial t} &+ \int_{0-}^{\infty} [f(t+v,j) - f(t,j)]\, \mu_{jj}^{(0)}\{dv\} \\ &+ \sum_{k>j} \int_{0-}^{\infty} [f(t+v,k) - f(t,j)]\, \mu_{jk}\{dv\}\end{aligned} \tag{3.52}$$

where

$$\mu_{jk}\{dv\} = q_{jk}\epsilon_0\{dv\} + \sum_{l<j} q_{jl} B_{lk}^{j}\{dv\} \quad (k > j) \tag{3.53}$$

$$\mu_{jj}^{(0)}\{dv\} = \sum_{l<j} q_{jl} B_{lj}^{j}\{dv\} \quad (0 \leq v < \infty) \tag{3.54}$$

$$\mu_{jj}^{(0)}\{+\infty\} = \sum_{l<j} q_{jl} P_l\{T_j = \infty\}, \tag{3.55}$$

q_{jk} $(k \neq j)$ are the transition rates for the process X, ϵ_0 is a distribution measure concentrated at the origin, B_{lk}^{j} are given by

$$B_{lk}^{j}\{A\} = P_l\{T_j \in A, X(T_j) = k\} \tag{3.56}$$

for $l < j \leq k$, with

$$T_j = \inf\{t \geq 0 : X(t) \geq j\}. \tag{3.57}$$

We note from (3.52) that (Y, J) is a Markov subordinator whose infinitesimal generator has the form (3.30) with the subordinator $X_{n+1}^{(1)}$ being a compound Poisson process with unit drift.

Now let

$$\zeta = \{(t, l) \in \mathbf{R}_+ \times \mathbf{N}_+ : M(t) = X(t) = l\}. \tag{3.58}$$

Proceeding as in Example 3.3 we see that ζ is a semiregenerative set. An inspection of the sample path of the process X shows that ζ is the range of the Markov subordinator (Y, J).

3.5 Semiregenerative Phenomena With Finite Label Set

For the semiregenerative phenomenon constructed in the last section we now suppose that E is a finite set. In this case $L = \infty$ a.s. and $P_{jk}(t)$ is given by (3.41). For $\theta > 0$ let

$$\hat{P}_{jk}(\theta) = \int_0^\infty \exp(-\theta t) P_{jk}(t)\, dt \tag{3.59}$$

and $\hat{P}(\theta) = (\hat{P}_{jk}(\theta))$. We have then the following.

Theorem 3.5. *For the semiregenerative phenomenon arising from the Markov-additive process (Y, J) of Sec. 3.4, with finite E, we have*

$$\left[\hat{P}(\theta)\right]^{-1} = R(\theta) \tag{3.60}$$

where $R(\theta) = (R_{jk}(\theta))$, with

$$R_{jj}(\theta) = \theta + \int_{0-}^{\infty^+} (1 - \exp(-\theta x))\mu_{jj}\{dx\} \tag{3.61}$$

$$R_{jk}(\theta) = -\int_{0-}^{\infty} \exp(-\theta x)\mu_{jk}\{dx\} \quad (j \neq k) \tag{3.62}$$

where μ_{jj} is a Lévy measure identical with $\mu_{jj}^{(0)}$ except that it has an additional weight λ_{jj} at infinity,

$$\mu_{jk}\{A\} = \lambda_{jk} F_{jk}\{A\} \tag{3.63}$$

and we note that

$$\sum_{k \neq j} \mu_{jk}\{\mathbf{R}_+\} \leq \mu_{jj}\{+\infty\}. \tag{3.64}$$

Proof. We first calculate the transform $r_j(\theta)$ of $p_j(t)$. We have

$$\int_0^\infty \exp(-\theta x) H_j\{s; dx\} = \exp(-s\phi_{jj}(\theta)) \quad (\theta > 0)$$

where

$$\phi_{jj}(\theta) = \theta + \int_{0-}^{\infty^+} (1 - \exp(-\theta x))\mu_{jj}^{(0)}\{dx\}.$$

From (3.35) we find that

$$\begin{aligned} r_j(\theta) &= \int_0^\infty \exp(-\theta t) p_j(t)\, dt \\ &= \int_0^\infty \exp(-\lambda_{jj} s - s\phi_{jj}(\theta))\, ds \\ &= \left[\theta + \int_{0-}^{\infty^+} (1 - \exp(-\theta x))\mu_{jj}\{dx\}\right]^{-1} \end{aligned} \tag{3.65}$$

and from (3.25) we find that

$$\hat{Q}_{jk}(\theta) = \int_{0-}^{\infty} \exp(-\theta t) Q_{jk}\{dt\} = \lambda_{jk} r_j(\theta) \hat{F}_{jk}(\theta) \tag{3.66}$$

where $\hat{F}_{jk}$ is the transform of F_{jk}.

Let us denote

$$\hat{U}_{jk}(\theta) = \int_{0-}^{\infty} \exp(-\theta t) U_{jk}\{dt\}, \tag{3.67}$$

$\hat{Q}(\theta) = (\hat{Q}_{jk}(\theta))$ and $\hat{U}(\theta) = (\hat{U}_{jk}(\theta))$. From (3.41) we find that

$$\hat{P}(\theta) = \hat{U}(\theta)(\delta_{jk} r_j(\theta))$$

where it is known from Markov renewal theory that $[\hat{U}(\theta)]^{-1} = I - \hat{Q}(\theta)$. Therefore

$$R(\theta) = \left[\hat{P}(\theta)\right]^{-1} = \left([r_j(\theta)]^{-1}\delta_{jk}\right)\left[I - \hat{Q}(\theta)\right].$$

Using (3.65)-(3.66) it is easily verified that the elements of the matrix on the right side of this last relation are given by (3.61)–(3.63). The inequality (3.64) follows from the fact that

$$\sum_{k \neq j} \lambda_{jk} F_{jk}\{\mathbf{R}_+\} \leq \sum_{k \neq j} \lambda_{jk} = \lambda_{jj} \leq \mu_{jj}\{+\infty\}.$$

□

We now ask whether semiregenerative phenomena with finite E can arise only in this manner (their semiregenerative sets corresponding to the range of a Markov-additive process of the type described in Sec. 3.4). In order to investigate this, we first prove the following result, which is essentially due to Kingman ([55], Theorem 5.2), who proved it in the setting of quasi-Markov chains. We shall only indicate the starting point of the proof.

Theorem 3.6. *Let Z be a standard semiregenerative phenomenon with finite E and $P_{jk}(t)$ defined by (3.9). Denote $\hat{P}(\theta)$ as in (3.59). Then*

$$\left[\hat{P}(\theta)\right]^{-1} = R(\theta) \tag{3.68}$$

where $R(\theta) = (R_{jk}(\theta))$, with $R_{jk}(\theta)$ given by (3.61)-(3.63), μ_{jk} $(j \neq k)$ being totally finite measures on $\mathbf{R}_+$ and μ_{jj} a Lévy measure on $[0, \infty]$. Moreover, the inequality (3.64) holds.

Proof. For each $h > 0$, let $Z'_h = \{Z'_{nl} = Z_{nh,l}, (n,l) \in \mathbf{N}_+ \times E\}$. Then Z'_h is a semirecurrent phenomena and, by Theorem 3.2,

$$P_{jk}(nh) = \sum_{r=1}^{\infty} q_{jk}^{(r)}(n)$$

with $q_{jk}^{(r)}(n)$ denoting the semirecurrence time distribution of Z', as defined in (3.14). This gives

$$\sum_{n=0}^{\infty} P_{jk}(nh)z^n = \delta_{jk} + \sum_{r=1}^{\infty}\sum_{n=1}^{\infty} q_{jk}^{(r)}(n)z^n$$

or

$$\begin{aligned}\left(\sum_{n=0}^{\infty} P_{jk}(nh)z^n\right) &= I + \sum_{r=1}^{\infty}\left(\sum_{n=1}^{\infty} q_{jk}(n)z^n\right)^r \\ &= \left[I - \left(\sum_{n=1}^{\infty} q_{jk}(n)z^n\right)\right]^{-1}.\end{aligned}$$

□

It turns out that the answer to the question raised above is affirmative. We prove this below.

Theorem 3.7. *Let Z be a standard semiregenerative phenomenon on $\mathcal{L} = \mathbf{R}_+ \times E$ with E finite. Then Z is equivalent to a phenomenon $\bar{Z}$ constructed as in Theorem 3.5, in the sense that Z and $\bar{Z}$ have the same $P_{jk}(t)$.*

Proof. We ignore the trivial case where all the measures $\mu_{jk}(j \neq k)$ are identically zero. Then from (3.64) we find that $\mu_{jj}\{+\infty\} > 0$. Define the probability measures F_{jk} by setting

$$\lambda_{jk}F_{jk}\{A\} = \frac{\mu_{jk}\{A\}}{\mu_{jj}\{+\infty\}} \quad (j,k \in E, j \neq k).$$

We construct an MAP on the state space $\mathbf{R}_+ \times E$ as in Sec. 3.4, with μ_{jj} and $\lambda_{jk}F_{jk}$. Let $\bar{Z}$ be the semiregenerative phenomenon obtained from this MAP and $\bar{P}_{jk}(t) = P\{\bar{Z}_{tk} = 1|\bar{Z}_{0j} = 1\}$. By Theorem 3.5 we have

$$\left[\hat{\bar{P}}(\theta)\right]^{-1} = \bar{R}(\theta)$$

where $\bar{R}(\theta) = (\bar{R}_{jk}(\theta))$, with $\bar{R}_{jk}(\theta) = R_{jk}(\theta)$. Thus $\hat{P}(\theta) = \hat{\bar{P}}(\theta)$. Since the $P_{jk}(t)$ are continuous functions, it follows that $P_{jk}(t) = \bar{P}_{jk}(t)$, as was required to be shown. □

Example (3.1) (continuation). For the continuous-time Markov chain J with finite state space E we find from (3.47) that

$$r_j(\theta) = (\theta + \lambda_{jj})^{-1},$$

so that the Lévy measure μ_{jj} is concentrated at ∞ with weight $\lambda_{jj} > 0$. Also the F_{jk} are concentrated at 0 with weight 1. Equality holds in (3.48). Theorem 3.5 gives

$$R_{jk}(\theta) = \begin{cases} \theta + \lambda_{jj} & \text{for } j = k \\ -\lambda_{jk} & \text{for } j \neq k \end{cases}$$

so that

$$R(\theta) = \left[\hat{P}(\theta)\right]^{-1} = \theta I - Q \tag{3.69}$$

where Q is the infinitesimal generator matrix of the chain. This is in agreement with the known result. By Theorem 3.7, the associated semiregenerative phenomenon is unique up to equivalence of $P_{jk}(t)$.

Chapter 4

Markov Random Walks: Fluctuation Theory and Wiener-Hopf Factorization

Markov random walks (MRWs) are discrete time versions of Markov-additive processes (MAPs). In this chapter, we present some basic theory of the MRW viewing it as a family of imbedded standard random walks. We consider a time-reversed version of the MRW which plays a key role in developing results analogous to those of the classical random walk. Specifically, by using the semirecurrent sets of the MRW and their time-reversed counterparts, we obtain: (a) The joint distributions of the extrema in terms of some measures of these sets; (b) A complete description of the fluctuation behavior of the nondegenerate MRW using the cardinalities of these sets; (c) A Wiener-Hopf factorization of the measures based on these sets; (d) The necessary and sufficient conditions for the renewal equation of the MRW to have an unique solution; and (e) The probabilistic behavior of exit times from a bounded interval of MRWs analogous to the Miller-Kemperman identity in classical random walks.

4.1 Introduction

Markov random walks are discrete time versions of MAPs as in Çinlar [28] and Chap. 2. Some authors used the terminology "random walks defined on a Markov chain" to refer to MRWs (see, e.g., Lotov and Orlova [61] and references cited there). Our terminology is self-explanatory: there is a family of random walks indexed by the state-space of an underlying Markov

[4]This chapter is an updated version of material included in the following two articles: N.U. Prabhu, L.C. Tang and Y. Zhu (1991), 'Some new results for the Markov random walk', *J. Math. Phys. Sci.* **25**, 5-6, pp. 635–663, and N.U. Prabhu (1994), 'Further results for semiregenerative phenomena', *Acta Appl. Math.* **34**, 1-2, pp. 213–223, whose contents are reproduced with permission from Kluwer Academic Publishers.

chain; and parallel to the well-known Markov renewal processes (MRPs).

Early pioneers in MRWs included Miller [68, 69], Presman [97], and Arjas and Speed [5–7], whose treatments are purely analytical, involving heavy use of operators, while later authors, Prabhu, Tang and Zhu [95, 112], adopted a more probabilistic approach. The latter probabilistic approach provides naturally the desired generality of the results presented in this chapter.

One of the central themes of this chapter is on the fluctuation theory of MRWs which was first studied by Newbould [75] for nondegenerate MRWs with a finite state ergodic Markov chain. Here, more general results for MRWs with countable state space are presented and a fundamental definition for degenerate MRW is given so that the probabilistic structure of MRWs can be firmly established. Together with the main result containing Wiener-Hopf factorization of the underlying transition measures, it provides better insights for models with Markov sources and facilitates potential applications.

In particular, MRWs have many applications to Markov-modulated queues and storage models, as will be illustrated in Chap. 7 and Chap. 9, respectively, similar to the applications of the random walk to classical queues and storage models (see, e.g., Prabhu [91]). Moreover, in recent years the Hidden Markov Model (HMM) has emerged as a plausible model for some interesting applications of MRWs in finance. For example, Siu [106] proposed the use of a Markov chain to model the evolution of the state of economy, and to model both the interest rate and volatility as random processes modulated by the underlying Markov chain. In addition, Elliot and Swishchuk [32] studied a Markov-modulated Brownian Market. They formulated the bond and risky asset price whose interest rate, drift and volatility terms are similarly modulated by an underlying chain. These are clearly connected to MRWs although the key interest in HMMs is in state estimation, which is of statistical concern, while that of MRWs is the fluctuation theory, which is probabilistic in nature. Indeed, the potential applications of MRWs in finance have been noted by Tang [112] (see Sec. 5.4). This connection between MRWs and HMMs will be clearer when their definitions are given later.

In Sec. 4.2, we define an MRW $\{(S_n, J_n),\, n \geq 0\}$ as a Markov process with the first component S_n having an additive property, J_n being a Markov chain on a countable state space E. We observe at the outset that the MRW is a family of imbedded (classical) random walks. This enables us in Sec. 4.3 to investigate the problem of degeneracy of the MRW in a broader framework and obtain more significant results than those obtained by Newbould [75].

We present the structure of time-reversed MRW and introduce the extrema and semirecurrent sets of the MRW in Sec. 4.4. In particular, we establish a simple yet useful result (Lemma 4.3) connecting the probability measures associated with the given MRW and its time-reversed counterpart. A special case of this result has been derived by Asmussen ([10], Theorem 3.1) without recognizing it as such (see Lemma 4.5). For the MRW the semirecurrent sets correspond to the observed ranges of the Markov renewal processes given by the sequences of ascending and descending ladder heights. The joint distributions of the extrema are then derived in terms of measures of these semirecurrent sets.

In Sec. 4.5, the fluctuation theory of the MRW is developed without the assumption of existence of means. Our approach is based on the theory of semirecurrent phenomena, developed by Prabhu [88] and presented in Sec. 3.3. In the case of a nondegenerate MRW with an irreducible and persistent J, it turns out that its fluctuation behavior is exactly the same as in the classical random walk (Theorem 4.8). This is an extension of a result obtained by Newbould [75] for the special case of a finite state ergodic J. For a degenerate MRW the situation is very different; here the process oscillates between (finite or infinite) bounds, except in the trivial case where $S_1 = S_2 = \cdots = 0$ for any initial state $J_0 = j$ (Theorem 4.3). We believe that our results concerning the fluctuation behavior of the MRW are much more comprehensive than others in the literature.

Then, in Sec. 4.6, we consider the case when the means exist and obtain some limit results and characterize the fluctuation behavior of MRWs using these results. Next, we look in Sec. 4.7 at a renewal equation for the MRW and establish the necessary and sufficient conditions for the equation to have an unique solution which are related to the fluctuation of MRW. We also provide a probabilistic proof of a matrix identity for the MRW analogous to the Miller-Kemperman identity in classical random walk.

In Sec. 4.8, we obtain a Wiener-Hopf factorization of the MRW in terms of positive Borel measures based on the associated semirecurrent sets. It turns out that this factorization is really a property of these measures. We show that the Fourier transforms of these measures, if necessary, can be derived and lead to the standard form of the factorization more familiar in the literature, expressed in terms of the transforms of ladder heights. The earliest result of this type is due to Miller [69] and the one demonstrated here is due to Presman [97]. Arjas and Speed [6, 7] derived various results for such factorizations using heavily analytical tools involving operators. Our approach is probabilistic and brings out the essential simplicity of the factorization.

Finally, we study in Sec. 4.9 the problem of exit of the MRW from a bounded interval, the classical version of which occurs in sequential analysis. For the nondegenerate MRW with J persistent nonnull, we obtain results analogous to those in sequential analysis concerning the exit time.

4.2 Definitions and Basic Properties

Suppose that we are given a probability space $(\Omega, \mathcal{F}, P)$. We denote $\mathbf{R} = (-\infty, \infty)$ and $E =$ countable set.

Definition 4.1. A Markov random walk $(S, J) = \{(S_n, J_n), n \geq 0\}$ is a Markov process on the state space $\mathbf{R} \times E$ whose transition distribution measure is given by

$$\begin{aligned} &P\{(S_{m+n}, J_{m+n}) \in A \times \{k\} | (S_m, J_m) = (x, j)\} \\ &\quad = P\{(S_{m+n} - S_m, J_{m+n}) \in (A - x) \times \{k\} | J_m = j\} \end{aligned} \tag{4.1}$$

for all $j, k \in E$ and Borel subset A of $\mathbf{R}$.

We shall only consider the time-homogeneous case where the second probability in (4.1) does not depend on m. We denote this probability as $Q_{jk}^{(n)}\{A - x\}$, so that

$$Q_{jk}^{(n)}\{A\} = P\{(S_n, J_n) \in A \times \{k\} | J_0 = j\}. \tag{4.2}$$

It is seen that the transition distribution measure Q satisfies the conditions

$$Q_{jk}^{(n)}\{A\} \geq 0, \qquad \sum_{k \in E} Q_{jk}^{(n)}\{\mathbf{R}\} = 1 \tag{4.3}$$

the Chapman-Kolmogorov equations are given by

$$Q_{jk}^{(m+n)}\{A\} = \sum_{l \in E} \int_{-\infty}^{\infty} Q_{jl}^{(m)}\{dx\} Q_{lk}^{(n)}\{A - x\} \quad (m, n \geq 0) \tag{4.4}$$

where

$$Q_{jk}^{(0)}\{A\} = \begin{cases} 1 & \text{if } j = k,\ 0 \in A \\ 0 & \text{otherwise} \end{cases} \tag{4.5}$$

and the initial measure is given by

$$P\{(S_0, J_0) \in \{A\} \times \{j\}\} = \begin{cases} a_j & \text{if } 0 \in A \\ 0 & \text{otherwise.} \end{cases} \tag{4.6}$$

From (4.2) and (4.6) the finite dimensional distributions of the process are defined, for $r \geq 1$, $0 = n_0 \leq n_1 \leq n_2 \leq \ldots \leq n_r$, $k_1, k_2, \ldots, k_r \in E$, and Borel subsets $A_1, A_2, \ldots, A_r$ of $\mathbf{R}$, with $x_0 = 0$, by

$$P\left\{\bigcap_{t=1}^{r}\{(S_{n_t}, J_{n_t}) \in \{A_t\} \times \{k_t\}\}\right\}$$
$$= \sum_{k_0 \in E} a_{k_0} \int_{\{x_t \in A_t\ (1 \leq t \leq r)\}} \prod_{t=1}^{r} Q_{k_{t-1}k_t}^{(n_t - n_{t-1})}(dx_t - x_{t-1}). \tag{4.7}$$

If the distribution of $\{S_n\}$ is concentrated on $[0, \infty)$, then the MRW defined above reduces to an MRP. On the other hand, the MRW is the discrete time analogue of MAP. From (4.1) and (4.2) we see that the marginal process $J = \{J_n,\ n \geq 0\}$ is a Markov chain on E with transition probabilities given by

$$P_{jk}^{(n)} = P\{J_n = k | J_0 = j\} = Q_{jk}^{(n)}\{\mathbf{R}\}. \tag{4.8}$$

We shall denote the conditional probabilities and expectations given $J_0 = j$ as P_j and E_j, respectively. As usual, we write Q_{jk} for $Q_{jk}^{(1)}$.

If the support of the observations of a Markov-modulated experiment is a collection of objects, typically represented by alphabets or symbols, then the random experiment leads to an Hidden Markov Model (HMM). These originated from interests in statistical methods for phenomena with Markov sources, with the source being frequently non-observable. A simple illustration of an HMM arises from a coin tossing process in which the observed states are heads (H) and tails (T) and the hidden Markov chain $\{L_n\}$ evolving on $E = \{F, f\}$ determines which coin to toss: fair (F) or foul (f). Let Z_n denote the coin side observed at the nth coin toss. If the different coin tossings are conditionally independent given $\{L_n\}$, then $\{(Z_n, L_n)\}$ is an HMM. Suppose there is a "reward" of losing 1 for "T" and winning \$1 for "H". At the nth play, the cumulative reward is $R_n = \sum_{j=1}^{n} I_n$, where $I_n = \mathbf{1}_{\{Z_n = H\}} - \mathbf{1}_{\{Z_n = T\}}$, defines a random walk modulated by $\{Z_n\}$. It is then clear that $\{(R_n, L_n)\}$ is an MRW and R_n denotes the difference between the numbers of heads and tails observed in the first n coin tosses.

Remark 4.1. The key difference between MRW and HMM lies in the state space endowed. The state spaces defined on HMMs can be arbitrary alphabets while that of an MRW has state space defined on $\mathbf{R}$. Hence, HMMs with alphabets not belonging to $\mathbf{R}$ cannot be MRWs. From the above illustration, however, it is conceivable that suitable mapping can be defined on an HMM so that the resulting process becomes a MRW. In addition, a general MRW modulated by a Markov chain with countable state space cannot be an HMM as the latter has finite state space.

To illustrate further the structure of the MRW in relation to those of Markov chain and random walk, it is useful to consider random walks imbedded in the MRW (S, J) on the state space $\mathbf{R} \times E$ as follows. We denote the successive hitting times of state $j \in E$ by $\tau_0^j = 0$ and

$$\tau_r^j = \min\{n > \tau_{r-1}^j : (S_n, J_n) \in \{\mathbf{R}\} \times \{j\}\} \quad (r \geq 1) \tag{4.9}$$

and the number of such hits (visits) as

$$N_n^j = \max\{r : \tau_r^j \leq n\}. \tag{4.10}$$

Also on $\tau_r^j < \infty$ we define

$$S_r^j = S_{\tau_r^j} \quad (r \geq 0). \tag{4.11}$$

We note that although the processes $\{S^j_{N_n^j}, n \geq 0\}$ and $\{S_r^j, r \geq 0\}$ have different index sets which affect their evolution, their observed ranges remain identical. In Chap. 5, we shall present the limit behavior of these processes and relate it to that of the MRW. First we make a simple observation arising from the properties of the imbedded processes.

Proposition 4.1. *$\{(\tau_r^j, S_r^j), r \geq 0\}$ is a random walk on the state space* $\mathbf{N}_+ \times \mathbf{R}$.

Proof. The sequence $\{\tau_r^j\}$ constitutes an imbedding renewal process, and the result follows from the strong Markov property of (S, J). □

We need a classification of the set E; this is essentially based on the recurrence properties of the marginal chain J, but it seems more appropriate to consider the two dimensional random walk $\{\tau_r^j, S_r^j\}$ rather than $\{\tau_r^j\}$.

Definition 4.2. State $j \in E$ is persistent if $\{\tau_1^j, S_1^j\}$ has a proper distribution, i.e.

$$P_j\{\tau_1^j < \infty, |S_1^j| < \infty\} = 1 \tag{4.12}$$

and transient otherwise.

Proposition 4.2. *Given $j \in E$, we have the following.*

(a) *If j is transient, then the random walk $\{\tau_r^j, S_r^j\}$ is terminating.*

(b) *If j is persistent and $P_j\{S_1^j = 0\} < 1$, then $\{S_r^j\}$ either (i) drifts to $+\infty$; (ii) drifts to $-\infty$; or (iii) oscillates a.s.*

Proof. (a). If j is transient then $\{\tau_r^j\}$ is a terminating renewal process.

(b). It is clear that the marginal process $\{S_r^j\}$ is a random walk and the result follows from the fact that any random walk that is nondegenerate at zero belongs to one of the 3 types indicated. □

To obtain a classification of the MRW similar to that of the classical random walk, we need to consider the case where S_n neither drifts nor oscillates. This special case seems to have been overlooked by Asmussen [9] who gave a classification of $\{S_n\}$ under the conditions $E_j(|X_1|) < \infty$, for all $j \in E$, with J persistent nonnull. Here is a counterexample that motivates the discussions in the next section.

Example 4.1. Let $E = \{-1, 0, 1\}$,

$$F_j\{A\} = \begin{cases} 1 & \text{if } j \in A \\ 0 & \text{otherwise} \end{cases}$$

for $j \in E$, and

$$\mathbf{Q}\{A\} = \begin{bmatrix} 0 & F_{-1}\{A\} & 0 \\ 0 & \frac{1}{2}F_0\{A\} & \frac{1}{2}F_0\{A\} \\ F_1\{A\} & 0 & 0 \end{bmatrix}.$$

Note that J is persistent nonnull with probability transition matrix

$$\mathbf{P} = \begin{bmatrix} 0 & 1 & 0 \\ 0 & 1/2 & 1/2 \\ 1 & 0 & 0 \end{bmatrix}.$$

Then it is clear that the observed range of $\{S_n\}$ is $\{-1, 0, 1\}$. This shows that this MRW does not fluctuate as in the classical random walk.

4.3 Degenerate Markov Random Walks

In the case of a standard random walk, degeneracy is the trivial case where $P(S_1 = 0) = 1$. In the case of the MRW, the conditions for degeneracy were investigated by Presman [97] and Newbould [75] who restricted J to be ergodic and to have finite state space. Here we shall start with the following definition.

Definition 4.3. The MRW is said to be degenerate at $j \in E$ if $\{S_r^j\}$ is degenerate at zero, i.e.

$$P_j(S_1^j = 0) = 1. \tag{4.13}$$

According to this definition j is persistent whenever the MRW is degenerate at j, since S_r^j is defined only if $\tau_r^j < \infty$. We shall show that degeneracy is a class property which will lead us to a natural way of defining degeneracy for the MRW.

Theorem 4.1. *Suppose that the MRW is degenerate at j and $Q_{jk}^{(n)}\{\mathbf{R}\} > 0$ for some n. Then the MRW is also degenerate at $k \in E$. In particular, if J is irreducible, then either all the imbedding random walks are nondegenerate or all are degenerate.*

Proof. As noted above if (4.13) holds for $j \in E$, then j is persistent. Given that $Q_{jk}^{(n)}\{\mathbf{R}\} > 0$ for some n, k is also persistent, from the theory of Markov chains. Thus, to prove that the MRW is degenerate at k it suffices to show that $P_k\{S_1^k = 0\} = 1$. On $\{J_0 = k\}$, let $\sigma_0^{jk} = 0$,

$$\tau_r^{kj} = \min\{n > \sigma_{r-1}^{jk} : J_n = j\} \quad (r \geq 1)$$

$$\sigma_r^{jk} = \min\{n > \tau_r^{kj} : J_n = k\} \quad (r \geq 1).$$

Since both j and k are persistent with $Q_{jk}^{(n)}\{\mathbf{R}\} > 0$, we have

$$\sigma_r^{jk} < \infty \text{ a.s.} \quad \text{and} \quad \tau_r^{kj} < \infty \text{ a.s.} \quad (r \geq 1).$$

It follows from the strong Markov property that

$$\left\{S_{\sigma_r^{jk}} - S_{\tau_r^{kj}}\right\}_{r\geq 1} \overset{\text{iid}}{\sim} \left\{S_{\sigma_1^{jk}} - S_{\tau_1^{kj}}\right\}$$

$$\left\{S_{\tau_r^{kj}} - S_{\sigma_{r-1}^{jk}}\right\}_{r\geq 1} \overset{\text{iid}}{\sim} \left\{S_{\tau_1^{kj}} - S_{\sigma_0^{jk}}\right\}$$

where $\overset{\text{iid}}{\sim}$ stands for independent and identically distributed, and the sequences $\{S_{\sigma_r^{jk}} - S_{\tau_r^{kj}}\}_{r\geq 1}$ and $\{S_{\tau_r^{kj}} - S_{\sigma_{r-1}^{jk}}\}_{r\geq 1}$ are independent. This in turn implies that

$$\left\{S_{\sigma_r^{jk}} - S_{\sigma_{r-1}^{jk}}\right\}_{r\geq 1} \overset{\text{d}}{=} \left\{S_{\tau_r^{kj}} - S_{\tau_{r-1}^{kj}}\right\}_{r\geq 2}$$

where $\stackrel{d}{=}$ denotes equality in distribution. Therefore, if j satisfies (4.13), then

$$P_k\{S_{\tau_r^{kj}} - S_{\tau_{r-1}^{kj}} = 0\} = 1 \quad (r \geq 2)$$

would imply that

$$P_k\{S_{\sigma_r^{jk}} - S_{\sigma_{r-1}^{jk}} = 0\} = 1 \quad (r \geq 1)$$

so that $P_k\{S_1^k = 0\} = 1$.

It follows from irreducibility of the J-chain that if (4.13) holds for some $j \in E$, it holds for all $j \in E$. □

Motivated by the previous theorem, it seems reasonable to define degeneracy of the MRW as follows.

Definition 4.4. The MRW is degenerate if all the imbedded random walks are degenerate, i.e., (4.13) holds for all $j \in E$.

Note that this definition would imply that for a degenerate MRW, J is a persistent chain. We have the following simple but useful result.

Lemma 4.1. *If J persistent and the MRW is uniformly bounded, then it is degenerate.*

Proof. If the MRW is uniformly bounded, then each imbedded random walk $\{S_r^j\}$ is also uniformly bounded since $\tau_r^j < \infty$ a.s. This means that all these imbedded random walks are degenerate so that the MRW is degenerate. □

The following theorem reveals the full implication of the term degeneracy, namely, that in a degenerate MRW, the increment between any transition $j \to k$ in J is deterministic, depending on j and k.

Theorem 4.2. *Suppose that J is persistent. Then the MRW is degenerate iff there exist finite constants $\{b_j\}_{j \in E}$ having the property that $Q_{jk}^{(n)}\{\cdot\}$ is concentrated on $\{b_k - b_j\}$ whenever $P_{jk}^{(n)} = Q_{jk}^{(n)}\{\mathbf{R}\} > 0$.*

Proof. (i). Suppose there exist constants b_j with the stated properties. Then since, for $j \in E$, $Q_{jj}^{(n)}\{0\} = P_{jj}^{(n)}$ for each n such that $P_{jj}^{(n)} > 0$, and j is persistent, (4.13) holds, i.e., the MRW is degenerate at j. Since J is irreducible, in view of Theorem 4.1 this implies that the MRW is degenerate.

(ii). By the definition of degeneracy, we have

$$S_{\tau_r^j} - S_{\tau_1^j} = 0 \;\; a.s. \quad (r \geq 1, j \in E)$$

regardless of J_0. Given $j, k \in E$, the previous relation implies that for any l, m, and n such that $\{J_l = j, J_{l+m} = k, J_{l+m+n} = j\}$ we must have $S_l = S_{l+m+n} = S_{\tau_1^j}$ and $S_{l+m} = S_{\tau_1^k}$ a.s. Accordingly,

$$S_{l+m} - S_l = S_{\tau_1^k} - S_{\tau_1^j} \ \ a.s.$$

$$S_{l+m+n} - S_{l+m} = S_{\tau_1^j} - S_{\tau_1^k} \ \ a.s.$$

$$S_{l+m+n} - S_l = 0 \ \ a.s.$$

As the random variables $S_{l+m} - S_l$ and $S_{l+m+n} - S_{l+m}$ are conditionally independent given $\{J(l) = j, J(l+m) = k, J(l+m+n) = j\}$ and their sum is the constant zero, a.s., they are necessarily equal to symmetric real constants, a.s. As a result, we conclude that there exists a real constant b_{jk} such that

$$S_{\tau_1^k} - S_{\tau_1^j} = b_{jk} = -[S_{\tau_1^j} - S_{\tau_1^k}] \ \ a.s.$$

This shows that the constants b_{jk} are anti-symmetric and, in particular, $b_{jj} = 0\,(j \in E)$. A similar reasoning leads to the conclusition that

$$b_{ik} = b_{ij} + b_{jk} \quad (i, j, k \in E).$$

Now, choosing $i \in E$ and a real constant b_i arbitrarily and letting

$$b_k = b_i + b_{ik} \quad (k \in E)$$

we conclude that

$$b_k - b_j = (b_i + b_{ik}) - (b_i + b_{ij}) = b_{ik} - b_{ij} = b_{jk} \quad (j, k \in E).$$

Thus, we conclude that $Q_{jk}^{(n)}\{\cdot\}$ is concentrated on $\{b_k - b_j\}$ whenever $P_{jk}^{(n)} = Q_{jk}^{(n)}\{\mathbf{R}\} > 0$, as wanted. □

The fluctuation aspect of degenerate MRWs can be briefly summarized as follows.

Theorem 4.3. *For a degenerate MRW we have P_j-a.s. $(j \in E)$*

$$\inf_k (b_k - b_j) = \liminf_n S_n \leq 0 \leq \limsup_n S_n = \sup_k (b_k - b_j). \tag{4.14}$$

If, in addition, E is finite then it is uniformly bounded, i.e., P_j-a.s.

$$-\infty < \min_k (b_k - b_j) = \liminf_n S_n \leq 0 \leq \limsup_n S_n = \max_k (b_k - b_j) < +\infty.$$

Proof. By definition of degeneracy, it follows that J is persistent, so that

$$P_j\{\tau_r^k < \infty,\, r = 1, 2, \ldots\} = 1 \quad (j, k \in E).$$

Thus, as (regardless of J_0) J visits infinitely often any of its states and, from Theorem 4.2,

$$P_j\{S_{\tau_r^k} = b_k - b_j\} = 1 \quad (r \geq 1,\, j, k \in E)$$

it follows that P_j-a.s. $(j \in E)$

$$\liminf S_n = \inf_{k \in E}(b_k - b_j) \quad \text{and} \quad \limsup S_n = \sup_{k \in E}(b_k - b_j).$$

By making $k = j$, we obtain

$$\inf_{k \in E}(b_k - b_j) \leq 0 \leq \sup_{k \in E}(b_k - b_j)$$

and the infimum and supremum above are finite and equal a minimum and a maximum, respectively, case E is finite. □

Remark 4.2. It should be noted that for a degenerate MRW the only possible limit of $\{S_n\}$, if it exists, is zero. This corresponds to the case where

$$P_j\{S_0 = S_1 = S_2 = \cdots = 0\} = 1 \qquad (j \in E). \tag{4.15}$$

Otherwise, the MRW "oscillates."

To see that finite E is necessary to ensure boundedness, consider the following degenerate MRW with $E = \{1, 2, 3, \ldots\}$.

Example 4.2. For even j,

$$Q_{j,j+1}\{A\} = \begin{cases} 1 & \text{if } \frac{1}{j} \in A \\ 0 & \text{otherwise.} \end{cases}$$

For odd j,

$$Q_{j,j+1}\{A\} = \begin{cases} \frac{1}{2} & \text{if } \frac{1}{j} \in A \\ 0 & \text{otherwise} \end{cases}$$

and

$$Q_{j1}\{A\} = \begin{cases} \frac{1}{2} & \text{if } -\sum_{k=1}^{j-1} \frac{1}{k} \in A \\ 0 & \text{otherwise.} \end{cases}$$

The transition matrix for the J-chain is

$$\mathbf{P} = \begin{bmatrix} \frac{1}{2} & \frac{1}{2} & 0 & 0 & 0 & 0 & \ldots \\ 0 & 0 & 1 & 0 & 0 & 0 & \ldots \\ \frac{1}{2} & 0 & 0 & \frac{1}{2} & 0 & 0 & \ldots \\ 0 & 0 & 0 & 0 & 1 & 0 & \ldots \\ \ldots & \ldots & \ldots & \ldots & \ldots & \ldots & \ldots \end{bmatrix}$$

so that J is persistent nonnull and $\limsup S_n = +\infty$ as $\sum_{k=1}^{\infty} \frac{1}{k}$ diverges.

We conclude this section by the following simple but useful result.

Lemma 4.2. *If an MRW with J persistent is uniformly bounded then it is degenerate.*

Proof. If the MRW is uniformly bounded, then each imbedded random walk is also uniformly bounded since $\tau_r^j < \infty$ a.s., for all $j \in E$. This means that all imbedded random walks are degenerate so that the MRW is degenerate. □

In view of Example 4.2, the converse of this lemma need not be true .

4.4 Time-Reversed MRW, Extrema, and Semirecurrent Sets

In this section we obtain some general results concerning the minimum and maximum functionals and the corresponding semirecurrent sets. Arndt [8] has obtained some results on the special case where the transforms of the increments can be represented as fractions of 2 polynomials. Throughout this section, we shall assume that J is irreducible and persistent. As in the classical setup, we denote

$$m_n = \min_{0 \leq k \leq n} S_k \quad \text{and} \quad M_n = \max_{0 \leq k \leq n} S_k \quad (n \geq 0). \tag{4.16}$$

We note that $m_n \leq 0$ and $M_n \geq 0$. Also $\{m_n\}$ is a nonincreasing sequence while $\{M_n\}$ is a non-decreasing sequence. Therefore, by monotone convergence theorem

$$m_n \to m \geq -\infty \quad \text{and} \quad M_n \to M \leq +\infty, \; a.s. \tag{4.17}$$

In the classical case where the increments are i.i.d., the distributions of these extrema are related in the sense that $(S_n - m_n, -m_n) \stackrel{\text{d}}{=} (M_n, M_n - S_n)$. This is, however, not true for the MRW in the presence of the modulating Markov chain J where the increments are no longer exchangeable.

Here, we consider a time-reversed version of the MRW and establish the relationship between its transition probability measure and that of its time-reversed counterpart. This is a fundamental result that connects the joint distributions of extrema to functions of measures for semirecurrent sets, which in turn leads to the Wiener-Hopf Factorization for MRW. The proof presented here invokes the Dynkin $\pi - \lambda$ Theorem which can be found, e.g., in [14, 98]. We first give the following definition.

Definition 4.5. $\{(\widehat{S}_n, \widehat{J}_n),\ n \geq 0\}$ is defined to be a time-reversed MRW induced by (S, J) if its transition probability measure is given by

$$\widehat{Q}_{jk}\{A\} = \frac{\pi_k}{\pi_j} Q_{kj}\{A\} \qquad (j, k \in E) \tag{4.18}$$

for any Borel subset A of $\mathbf{R}$, where $Q_{kj}\{\cdot\}$ is the transition probability measure of (S, J) and $\{\pi_i\}$ is a probability distribution on E.

By setting $A = \mathbf{R}$ in the above equation, we see that

$$\pi_j \widehat{P}_{jk} = \pi_k P_{kj} \qquad (j, k \in E)$$

which shows that $\{\pi_i\}$ is necessarily the stationary distribution of both J and $\widehat{J}$.

Henceforth, $\widehat{(\cdot)}$ will denote the counterpart of $(\cdot)$ for the time-reversed MRW. In addition, we let $X_r = S_r - S_{r-1}$ $(r \geq 1)$ denote the increments of the MRW and $\mathcal{F}_n$ denote the σ-field generated by $(X_1, X_2, \ldots, X_n)$. Moreover, given a set $\mathcal{A}_n \in \mathcal{F}_n$, we let

$$\widehat{\mathcal{A}}_n = \{\omega : (X_n(\omega), X_{n-1}(\omega), \ldots, X_1(\omega)) \in \mathcal{A}_n\}$$

so that $\widehat{\mathcal{A}}_n$ is the set corresponding to $\mathcal{A}_n$ for the time-reversed MRW.

The following result relates the given MRW to its time-reversed counterpart; in particular, it will lead us to the n-step transition probability measure for the time-reversed MRW.

Lemma 4.3. *For all $j, k \in E$ and $\mathcal{A}_n \in \mathcal{F}_n$, we have*

$$P_k\{\mathcal{A}_n; J_n = j\} = \frac{\pi_j}{\pi_k} P_j\{\widehat{\mathcal{A}}_n; \widehat{J}_n = k\}. \tag{4.19}$$

Proof. We first consider the special case where

$$\mathcal{A}_n = \{\omega : X_r(\omega) \in A_r,\ r = 1, \ldots, n\} \tag{4.20}$$

with the sets A_r being Borel subsets of $\mathbf{R}$. For $j, k \in E$, we have

$$\begin{aligned}
P_k\{\mathcal{A}_n; J_n = j\} &= P_k\{X_r \in A_r, r = 1, \ldots, n; J_n = j\} \\
&= \sum_{i_1 \in E} \cdots \sum_{i_{n-1} \in E} \int_{A_n} \cdots \int_{A_1} Q_{ki_1}\{dx_1\} \cdots Q_{i_{n-1}j}\{dx_n\} \\
&= \sum_{i_1 \in E} \cdots \sum_{i_{n-1} \in E} \int_{A_1} \cdots \int_{A_n} \frac{\pi_j}{\pi_{i_{n-1}}} \widehat{Q}_{ji_{n-1}}\{dx_n\} \cdots \frac{\pi_{i_1}}{\pi_k} \widehat{Q}_{i_1k}\{dx_1\} \\
&= \frac{\pi_j}{\pi_k} \sum_{i_1 \in E} \cdots \sum_{i_{n-1} \in E} \int_{A_1} \cdots \int_{A_n} \widehat{Q}_{ji_{n-1}}\{dx_n\} \cdots \widehat{Q}_{i_1k}\{dx_1\} \\
&= \frac{\pi_j}{\pi_k} P_j\{\widehat{\mathcal{A}}_n; \widehat{J}_n = k\}.
\end{aligned}$$

The collection of sets for which the above equality holds is a $\lambda-$system and we have shown (4.19) for the collection of sets $\mathcal{A}_n$ of the form (4.20), which is a $\pi-$system. The Dynkin π-λ theorem implies that the desired identity holds for all $\mathcal{A}_n \in \mathcal{F}_n$, as wanted. □

As a special case of Lemma 4.3 we obtain the n-step transition probability measure of the time-reversed MRW as follows. Other special cases will be treated later in Lemma 4.5 and Lemma 4.9.

Theorem 4.4. *For all $j, k \in E$ and Borel subset A of* $\mathbf{R}$,

$$\widehat{Q}_{jk}^{(n)}\{A\} = \frac{\pi_k}{\pi_j} Q_{kj}^{(n)}\{A\} \qquad (n \geq 1). \tag{4.21}$$

Proof. In view of Lemma 4.3, we only need to define a correct form of $\mathcal{A}_n$ and find the corresponding $\widehat{\mathcal{A}}_n$. For $n \geq 1$ and a Borel subset A of $\mathbf{R}$, let $\mathcal{A}_n = \{\omega : S_n \in A\}$. Since

$$\mathcal{A}_n = \{\omega : X_1 + \cdots + X_n \in A\} = \{\omega : X_n + \cdots + X_1 \in A\} = \widehat{\mathcal{A}}_n$$

we have, in view of (4.19),

$$\widehat{Q}_{jk}^{(n)}\{A\} = \frac{\pi_k}{\pi_j} Q_{kj}^{(n)}\{A\}$$

for all $j, k \in E$, as wanted. □

With the above result on time-reversed MRW, the distributions of the extremals and their time reversed counterparts have the following relationship.

Lemma 4.4. *For $x, y \geq 0$ and $n \geq 1$, we have*

$$\begin{aligned} P_k\{S_n - m_n \leq x, -m_n \leq y, J_n = j\} \\ = \frac{\pi_j}{\pi_k} P_j\{\widehat{M}_n \leq x, \widehat{M}_n - \widehat{S}_n \leq y, \widehat{J}_n = k\}. \end{aligned} \tag{4.22}$$

Proof. The result follows from the fact that, in view of Lemma 4.3, the probability on the right side of (4.22) is

$$\begin{aligned} &P_j\{\widehat{M}_n \leq x, \widehat{M}_n - \widehat{S}_n \leq y, \widehat{J}_n = k\} \\ &= P_j\left\{\max_{r\leq n} \widehat{S}_r \leq x, \max_{r\leq n}(\widehat{S}_r - \widehat{S}_n) \leq y, \widehat{J}_n = k\right\} \\ &= \frac{\pi_k}{\pi_j} P_k\left\{\max_{r\leq n}(S_n - S_{n-r}) \leq x, \max_{r\leq n}(-S_{n-r}) \leq y, J_n = j\right\} \\ &= \frac{\pi_k}{\pi_j} P_k\{S_n - m_n \leq x, -m_n \leq y, J_n = j\}. \end{aligned}$$

□

Associated with the extremum functionals are the semirecurrent sets defined as follows

$$\begin{cases} \zeta_+ = \{(n,j) : S_n = M_n, J_n = j, \rho_n = n\} \\ \zeta_- = \{(n,j) : S_n = m_n, J_n = j\} \end{cases} \tag{4.23}$$

where ρ_n is the first epoch at which the MRW attains the value M_n. We also observe that ζ_+ and ζ_- are sets with the following elements:

$$\begin{cases} \zeta_+ &= \{(T_0, J_0), (T_1, J_{T_1}), \dots\} \\ \zeta_- &= \{(\bar{T}_0, J_0), (\bar{T}_1, J_{\bar{T}_1}), \dots\} \end{cases} \tag{4.24}$$

where

$$\begin{cases} T_r &= \min\{n > T_{r-1} : S_n > S_{T_{r-1}}\} \\ \bar{T}_r &= \min\{n > \bar{T}_{r-1} : S_n \leq S_{\bar{T}_{r-1}}\} \end{cases} \tag{4.25}$$

with $T_0 = \bar{T}_0 = 0$. We then have the following.

Theorem 4.5. *The processes* $\{(S_{T_r}, T_r, J_{T_r})\}$ *and* $\{(-S_{\bar{T}_r}, \bar{T}_r, J_{\bar{T}_r})\}$ *are MRP on the state space* $\mathbf{R}_+ \times \mathbf{N}_+ \times E$.

Proof. Let $Y_p = (S_{T_p}, T_p, J_{T_p})$, $p \geq 0$. Since the T_p are stopping times for the MRW, we have the following by the strong Markov property. For $r \geq 1$, $0 = n_0 \leq n_1 \leq \dots \leq n_{r-1} \leq n$, $j_0, j_1, \dots, j_{r-1}, j \in E$, $x_1, x_2, \dots, x_{r-1} \in \mathbf{R}$, and a Borel subset A of $\mathbf{R}$, with $x_0 = 0$:

$$\begin{aligned} P\{Y_r &\in A \times \{n\} \times \{j\} | Y_m = (x_m, n_m, j_m),\ 0 \leq m \leq r-1\} \\ &= P\{Y_r \in A \times \{n\} \times \{j\} | Y_{r-1} = (x_{r-1}, n_{r-1}, j_{r-1})\} \\ &= P\{S_{T_r} - S_{T_{r-1}} \in \{A - x_{r-1}\}, T_r - T_{r-1} = n - n_{r-1}, J_{T_r} = j \\ &\qquad\qquad\qquad\qquad\qquad\qquad | J_{T_{r-1}} = j_{r-1}\} \\ &= P_{j_{r-1}}\{Y_1 \in \{A - x_{r-1}\} \times \{n - n_{r-1}\} \times \{j\}\} \end{aligned}$$

which shows that the process considered is an MRP. The proof for $\{-S_{\bar{T}_r}, \bar{T}_r, J_{\bar{T}_r}\}$ is similar. □

Remark 4.3. In view of Theorem 3.3 (see also Prabhu [88]), the sets ζ_+ and ζ_- are semirecurrent sets corresponding to the MRP $\{S_{T_r}, T_r, J_{T_r}\}$ and $\{(-S_{\bar{T}_r}, \bar{T}_r, J_{\bar{T}_r})\}$, respectively.

Remark 4.4. The marginal Markov chain J_T associated with the above MRP $\{(S_{T_k}, T_k, J_{T_k})\}$ does not necessarily inherit properties such as ergodicity and aperiodicity from the J-chain in the case where E is countably infinite. This can be easily seen from Example 4.2 that J_T has transition probability matrix

$$\begin{bmatrix} 0 & 1 & 0 & 0 & 0 & \cdots \\ 0 & 0 & 1 & 0 & 0 & \cdots \\ 0 & 0 & 0 & 1 & 0 & \cdots \\ \cdots & \cdots & \cdots & \cdots & \cdots & \cdots \end{bmatrix}.$$

Thus, J_T is transient although the J-chain is persistent. This implies that Arjas's results [5] do not hold for the case where E is countably infinite.

Connected to the sets ζ_+ and ζ_- are the following measures. Let, for $n \geq 0$, $j, k \in E$, and Borel subsets I_+ and I_- of $\mathbf{R}_+$ and $\mathbf{R}_- = (-\infty, 0]$, respectively,

$$u_{jk}^{(n)}(I_+) = P_j\{(n,k) \in \zeta_+, S_n \in I_+\} \tag{4.26}$$

$$v_{jk}^{(n)}(I_-) = P_j\{(n,k) \in \zeta_-, S_n \in I_-\} \tag{4.27}$$

so that, in particular, $u_{jk}^{(0)}(I_+) = \delta_{jk}\mathbf{1}_{\{0 \in I_+\}}$ and $v_{jk}^{(0)}(I_-) = \delta_{jk}\mathbf{1}_{\{0 \in I_-\}}$, and

$$u_{jk}(I_+) = \sum_{n=0}^{\infty} u_{jk}^{(n)}(I_+) \quad \text{and} \quad v_{jk}(I_-) = \sum_{n=0}^{\infty} v_{jk}^{(n)}(I_-). \tag{4.28}$$

It should be noted that these measures are all finite as $u_{jk}(\cdot)$ and $u_{jk}(\cdot)$ are Markov renewal measures of the MRP $\{(S_{T_k}, J_{T_k})\}$ and $\{(-S_{\bar{T}_k}, J_{\bar{T}_k})\}$, respectively. It also follows from the theory of MRP that $\sum_{k \in E} u_{jk}(I_+)$ and $\sum_{k \in E} v_{jk}(I_-)$ are finite.

Lemma 4.5. *For Borel subsets I_+ of $\mathbf{R}_+$ and I_- of $\mathbf{R}_-$, we have for $n \in \mathbf{N}_+$ and $j, k \in E$,*

$$\widehat{u}_{jk}^{(n)}(I_+) = \frac{\pi_k}{\pi_j} P_k\{\bar{T}_1 > n, S_n \in I_+, J_n = j\}; \tag{4.29}$$

$$\widehat{v}_{jk}^{(n)}(I_-) = \frac{\pi_k}{\pi_j} P_k\{T_1 > n, S_n \in I_-, J_n = j\}. \tag{4.30}$$

Proof. The result is trivial case $n = 0$. Moreover, for $n \geq 1$, we have from the definition

$$\begin{aligned} \widehat{u}_{jk}^{(n)}(I_+) &= P_j\{(n,k) \in \widehat{\zeta}_+, \widehat{S}_n \in I_+\} \\ &= P_j\{\widehat{S}_n > \widehat{S}_1, \widehat{S}_n > \widehat{S}_2, \ldots, \widehat{S}_n > \widehat{S}_{n-1}, \widehat{S}_n \in I_+, \widehat{J}_n = k\}. \end{aligned}$$

Using the result (4.19) in the last probability, we obtain

$$\begin{aligned}\widehat{u}_{jk}^{(n)}(I_+) &= \frac{\pi_k}{\pi_j} P_k\{S_n > S_n - S_{n-m}, 1 \le m \le n-1, S_n \in I_+, J_n = j\} \\ &= \frac{\pi_k}{\pi_j} P_k\{S_n > 0, S_{n-1} > 0, \dots, S_1 > 0, S_n \in I_+, J_n = j\} \\ &= \frac{\pi_k}{\pi_j} P_k\{\bar{T}_1 > n, S_n \in I_+, J_n = j\}.\end{aligned}$$

This proves (4.29) and the proof of (4.30) is similar. □

Moreover, we can express the joint distributions of the extremum functionals as functions of the measures $u_{jk}^{(n)}(\cdot)$, $v_{jk}^{(n)}(\cdot)$, $\widehat{u}_{jk}^{(n)}(\cdot)$, and $\widehat{v}_{jk}^{(n)}(\cdot)$. To easy the notation, we let, for $x, y \ge 0$: $u_{jk}^{(n)}(x) = u_{jk}^{(n)}([0,x])$, $v_{jk}^{(n)}(-y) = v_{jk}^{(n)}([-y,0])$, $\widehat{u}_{jk}^{(n)}(x) = \widehat{u}_{jk}^{(n)}([0,x])$, and $\widehat{v}_{jk}^{(n)}(-y) = \widehat{v}_{jk}^{(n)}([-y,0])$.

Theorem 4.6. *For $x, y \ge 0$, $n \in \mathbf{N}_+$, and $i, k \in E$, we have*

$$\begin{aligned} & P_i\{M_n \le x, M_n - S_n \le y, J_n = k\} \\ & \qquad = \pi_k \sum_{j\in E} \sum_{m=0}^{n} u_{ij}^{(m)}(x) \frac{1}{\pi_j} \widehat{v}_{kj}^{(n-m)}(-y). \qquad (4.31) \\ & P_k\{S_n - m_n \le x, -m_n \le y, J_n = i\} \\ & \qquad = \pi_i \sum_{j\in E} \sum_{m=0}^{n} \widehat{u}_{ij}^{(m)}(x) \frac{1}{\pi_j} v_{kj}^{(n-m)}(-y). \qquad (4.32)\end{aligned}$$

In particular, we have

$$P_i\{M_n \le x, J_n = k\} = \pi_k \sum_{j\in E} \sum_{m=0}^{n} u_{ij}^{(m)}(x) \frac{1}{\pi_j} \widehat{v}_{kj}^{(n-m)}(-\infty). \qquad (4.33)$$

$$P_k\{m_n \ge -y, J_n = i\} = \pi_i \sum_{j\in E} \sum_{m=0}^{n} \widehat{u}_{ij}^{(m)}(\infty) \frac{1}{\pi_j} v_{kj}^{(n-m)}(-y). \qquad (4.34)$$

Proof. Let $T(n) = \max\{r : T_r \le n\}$. Then

$$\begin{aligned} & P_i\{M_n \le x, M_n - S_n \le y, J_n = k\} \\ & \qquad = \sum_{j\in E} P_i\{M_n \le x, M_n - S_n \le y, J_{T_{T(n)}} = j, J_n = k\}\end{aligned}$$

and, for $j \in E$, with $Y_p = (S_{T_p}, T_p, J_{T_p})$,

$$
\begin{aligned}
&P_i\{M_n \le x, M_n - S_n \le y, J_{T_{T(n)}} = j, J_n = k\} \\
&= \sum_{r=0}^{n} P_i\{T(n) = r, M_n \le x, M_n - S_n \le y, J_{T_r} = j, J_n = k\} \\
&= \sum_{r=0}^{n} \sum_{m=r}^{n} P_i\{T_r = m, T_{r+1} > n, S_{T_r} \le x, S_{T_r} - S_n \le y, J_n = k, J_{T_r} = j\} \\
&= \sum_{m=0}^{n} \sum_{r=0}^{m} \int_0^x P_i\{T_{r+1} > n, S_{T_r} - S_n \le y, J_n = k | Y_r = (z, m, j)\} \\
&\qquad\qquad \cdot P_i\{T_r = m, S_{T_r} \in dz, J_{T_r} = j\} \\
&= \sum_{m=0}^{n} \sum_{r=0}^{m} P_i\{T_r = m, S_m \le x, J_m = j\} \\
&\qquad\qquad \cdot P_j\{T_1 > n - m, S_{n-m} \ge -y, J_{n-m} = k\} \\
&= \pi_k \sum_{m=0}^{n} u_{ij}^{(m)}(x) \frac{1}{\pi_j} \widehat{v}_{kj}^{(n-m)}(-y)
\end{aligned}
$$

where the last equality follows from (4.30). This completes the proof of (4.31). From Lemma 4.4 and (4.31), we have

$$
\begin{aligned}
P_k\{S_n - m_n \le x, -m_n \le y, J_n = i\}
&= \frac{\pi_i}{\pi_k} P_i\{\widehat{M}_n \le x, \widehat{M}_n - \widehat{S}_n \le y, \widehat{J}_n = k\} \\
&= \frac{\pi_i}{\pi_k} \sum_{j \in E} \sum_{m=0}^{n} \widehat{u}_{ij}^{(m)}(x) \frac{\pi_k}{\pi_j} v_{kj}^{(n-m)}(-y) \\
&= \pi_i \sum_{j \in E} \sum_{m=0}^{n} \widehat{u}_{ij}^{(m)}(x) \frac{1}{\pi_j} v_{kj}^{(n-m)}(-y)
\end{aligned}
$$

which gives (4.32). Letting $y \to \infty$ in (4.31), we obtain (4.33). Similarly letting $x \to \infty$ in (4.32), we get (4.34). □

4.5 Fluctuation Theory for MRW

In this section we give a classification of the additive component of a non-degenerate MRW which is parallel to that of a classical random walk. The classification will be based on the maximum and minimum functionals and the corresponding semirecurrent sets discussed in the previous section with a slight change in notation as follows. Let $\zeta_+(\widehat{\zeta}_+)$ and $\zeta_-(\widehat{\zeta}_-)$ denote respectively the weak ascending and weak descending ladder sets of the MRW (time-reversed MRW); thus, in particular

$$\zeta_+ = \{(n, j) : S_n = M_n, J_n = j\}. \tag{4.35}$$

Throughout this section, we shall assume that J is persistent.

Before we provide a classification of the MRW, we give a result concerning the limit behavior of S_n in relation to M and m which is identical to that of a classical random walk.

Theorem 4.7. *For a nondegenerate MRW with J persistent it is impossible for both M and m to be finite. Furthermore, we have the following.*

(a) *If $M_n \to +\infty$ and $m_n \to m > -\infty$ a.s., then $S_n \to +\infty$ a.s.*
(b) *If $M_n \to M < +\infty$ and $m_n \to -\infty$ a.s., then $S_n \to -\infty$ a.s.*
(c) *If $M_n \to +\infty$ and $m_n \to -\infty$ a.s., then*

$$-\infty = \liminf_n S_n < \limsup_n S_n = +\infty \quad a.s.$$

Proof. Since $\sup_{n\geq 1} |S_n| = \max\{M, -m\}$, by Lemma 4.1 the MRW cannot be bounded. Accordingly we need consider only cases (a)–(c).

(a). From the Markov property and the fact that $M_{T_r} = S_{T_r}$, we have, for $r \in \mathbf{N}_+$ and a Borel subset I of $\mathbf{R}_-$:

$$\begin{aligned} P_{J_0}\{\inf_{n\geq T_r} S_n - M_{T_r} \in I | S_m, J_m; 0 \leq m \leq T_r\} &\\ = P_{J_0}\{\inf_{n\geq T_r} S_n - S_{T_r} \in I | S_{T_r}, J_{T_r}\} &\\ = P_{J_{T_r}}\{\inf_{n\geq T_r} S_{n-T_r} \in I\} &\\ = P_{J_{T_r}}\{m \in I\}. & \end{aligned}$$

So for $a, b > 0$,

$$P_{J_0}\left\{\inf_{n\geq T_r} S_n > a\right\} \geq P_{J_0}\{M_{T_r} > a + b\} P_{J_{T_r}}\{m > -b\}.$$

Since $M_n \to +\infty$ and $m_n \to m > -\infty$ a.s., given positive a and ϵ, we can find some sufficiently large b and r so that

$$P_{J_0}\{M_{T_r} > a + b\} > 1 - \epsilon \quad \text{and} \quad P_{J_{T_r}}\{m > -b\} > 1 - \epsilon.$$

Therefore

$$P_{J_0}\left\{\inf_{n\geq T_r} S_n > a\right\} > (1-\epsilon)^2 > 1 - 2\epsilon,$$

which shows that $S_n \to +\infty$ a.s.

(b). This follows from (a) by symmetry.

(c). $M_{T_r} = S_{T_r} \to +\infty$ and $m_{\bar{T}_r} = S_{\bar{T}_r} \to -\infty$ give us two divergent sequences that prove our assertion. □

Theorem 4.8. *For a nondegenerate MRW with J persistent it is impossible for both ζ_+ and ζ_- to be terminating.*

Furthermore, we have the following.

(a) *ζ_- is terminating and ζ_+ is nonterminating*

$$\Leftrightarrow m > -\infty,\ M = +\infty \Leftrightarrow \lim_n S_n = +\infty.$$

(b) *ζ_- is nonterminating and ζ_+ is terminating*

$$\Leftrightarrow m = -\infty,\ M < +\infty \Leftrightarrow \lim_n S_n = -\infty.$$

(c) *ζ_- and ζ_+ are both nonterminating*

$$\Leftrightarrow m = -\infty,\ M = +\infty$$
$$\Leftrightarrow \liminf_n S_n = -\infty,\ \limsup_n S_n = +\infty.$$

Proof. We first show that for the given MRW

(i) ζ_+ is terminating iff $M < \infty$
(ii) ζ_- is terminating iff $m > -\infty$.

By symmetry it suffices to show (i). Note that

$$\zeta_+ \text{ is terminating} \Longrightarrow \exists N < \infty : S_n \le S_N, n \le N \text{ and } S_n < S_N, n > N$$
$$\Longrightarrow M = S_N = \sum_{i=1}^{N} X_i < \infty \text{ a.s.}$$

Conversely, suppose $M < \infty$. By Lemma 4.2 the MRW is unbounded so $m = -\infty$. But from Theorem 4.7

$$M < \infty,\ \ m = -\infty \Longrightarrow \lim_n S_n = -\infty \Longrightarrow \zeta_+ \text{ is terminating}$$

as there is a last epoch n for which $S_n \in \mathbf{R}_+$. Thus, (i) holds.

From (i) and (ii), we conclude that

- ζ_- is terminating and ζ_+ is nonterminating $\Leftrightarrow m > -\infty,\ M = +\infty$;
- ζ_- is nonterminating and ζ_+ is terminating $\Leftrightarrow m = -\infty,\ M < +\infty$;
- ζ_- and ζ_+ are both nonterminating $\Leftrightarrow m = -\infty,\ M = +\infty$.

The previous results are the first equivalences in (a)-(c). The second equivalences concerning S_n for these three cases follow from Theorem 4.7. □

Remark 4.5. To see the need for nondegeneracy in the above theorem, we make a slight modification on the transition measure Q in Example 4.2 to one which makes $\{S_n\}$ concentrate its values on $\sum \frac{1}{k^2}$, by replacing the fractions $1/k$ by $1/k^2$ in the definition of Q. In this case ζ_+ is nonterminating but $M < \infty$ since $\sum \frac{1}{k^2}$ converges.

Corollary 4.1. *Suppose that the MRW is nondegenerate with persistent J. Then we have*

(a) $P_j\{\widehat{T} = \infty\} = P_j\{\bar{T} = \infty\} = 0, \forall j$

$$\Leftrightarrow \liminf_n S_n = -\infty, \limsup_n S_n = +\infty;$$

(b) $P_j\{\widehat{T} = \infty\} = 0, \forall j,$ *and* $P_j\{\bar{T} = \infty\} > 0,$ *for some* j

$$\Leftrightarrow \lim_n S_n = +\infty;$$

(c) $P_j\{\widehat{T} = \infty\} > 0$ *for some* $j,$ *and* $P_j\{\bar{T} = \infty\} = 0, \forall j$

$$\Leftrightarrow \lim_n S_n = -\infty.$$

Proof. First, we observe that $S_n = \widehat{S}_n$ and so $\limsup_n S_n = \limsup_n \widehat{S}_n$. Hence, we can replace $\widehat{T}$ by T in (a)-(c). The results then follow from Theorem 4.8, where $P_j\{T = \infty\} = 0$ (> 0) corresponds to the case where ζ_+ is nonterminating (terminating) and $P_j\{\bar{T} = \infty\} = 0$ (> 0) corresponds to the case where ζ_- is nonterminating (terminating). □

4.6 The Case Where Means Exist

In this section we consider, for J irreducible, the case where

$$\sup_{j \in E} E_j|X_1| < \infty \tag{4.36}$$

where we recall that $X_r = S_r - S_{r-1}$ $(r \geq 1)$ are the increments of the MRW. With this assumption, we obtain some elementary results concerning the Cesáro limits of the additive component of a MRW and its imbedded random walks. We then give a classification of the MRW based on its mean. First, we have the following result.

Lemma 4.6. *If the condition (4.36) holds, then*

$$E_j(S_1^j) = \sum_{k \in E} {}^jP_{jk}^* E_k(X_1) \quad (j \in E) \tag{4.37}$$

where ${}^jP_{jk}^* = \sum_{m=0}^{\infty} P_j(J_m = k, \tau_1^j > m)$. *In the case where the stationary probability measure* $\{\pi_k, k \in E\}$ *of the marginal chain* J *exists, (4.37) reduces to*

$$E_j(S_1^j) = \sum_{k \in E} \frac{\pi_k E_k(X_1)}{\pi_j}. \tag{4.38}$$

Proof. For $j \in E$, we have

$$\begin{aligned} E_j(S_1^j) &= \sum_{m=1}^{\infty} E_j\left(X_m \mathbf{1}_{\{\tau_1^j \geq m\}}\right) \\ &= \sum_{m=1}^{\infty} \sum_{k \in E} E_j(X_m | J_{m-1} = k) P_j(J_{m-1} = k, \tau_1^j > m-1) \\ &= \sum_{k \in E} E_k(X_1) \sum_{m=0}^{\infty} P_j(J_m = k, \tau_1^j > m) \\ &= \sum_{k \in E} {}^jP_{jk}^* E_k(X_1). \end{aligned}$$

This leads to (4.37). Moreover, (4.38) follows from (4.37) using the fact that from Markov chain theory ${}^jP_{jk}^* = \frac{\pi_k}{\pi_j}$. □

Theorem 4.9. *Let* $j \in E$ *be persistent. If* $E_j(|S_1^j|) < \infty$ *then*

$$\lim_{n \to \infty} \frac{S_{N_n^j}^j}{n} = \frac{E_j(S_1^j)}{E_j(\tau_1^j)} < \infty \quad a.s. \tag{4.39}$$

If the stationary probability measure $\{\pi_k, k \in E\}$ *exists, (4.39) reduces to*

$$\lim_{n \to \infty} \frac{S_{N_n^j}^j}{n} = \sum_{k \in E} \pi_k E_k(X_1) \quad a.s. \tag{4.40}$$

Proof. From the elementary renewal theorem we have

$$\lim_{n \to \infty} \frac{N_n^j}{n} = \frac{1}{E_j(\tau_1^j)} \quad a.s.$$

where $E_j(\tau_1^j) \leq \infty$. Also, since j is persistent, $N_n^j \to \infty$ a.s. as $n \to \infty$, and so it follows from the Strong Law of Large Numbers that $S_r^j / r \to E_j(S_1^j)$ a.s. Therefore,

$$\frac{S_{N_n^j}^j}{N_n^j} \longrightarrow E_j(S_1^j) \quad a.s.$$

and

$$\frac{S^j_{N^j_n}}{n} = \frac{S^j_{N^j_n}}{N^j_n} \cdot \frac{N^j_n}{n} \longrightarrow \frac{E_j(S^j_1)}{E_j(\tau^j_1)} \quad a.s.$$

Moreover, (4.40) follows from (4.39) using the fact that if the stationary probability measure $\{\pi_k, k \in E\}$ exists, then (4.38) holds and $E_j(\tau^j_1) = 1/\pi_j$. □

With these preliminary results we are now ready to consider the additive component $\{S_n\}$.

Lemma 4.7. *For $j, k \in E$ and $n \in \mathbf{N}_+$, we have*

$$E_j[S_n; J_n = k] = \sum_{m=0}^{n-1} \sum_{l,l' \in E} P^{(m)}_{jl} E_l[X_1; J_1 = l'] P^{(n-1-m)}_{l'k} \tag{4.41}$$

in the sense that both sides of the equation are finite or infinite together.

Proof. We have

$$\begin{aligned} E_j[S_n; J_n = k] &= \sum_{m=0}^{n-1} E_j[X_{m+1}; J_n = k] \\ &= \sum_{m=0}^{n-1} \sum_{l,l' \in E} E_j[X_{m+1}; J_m = l, J_{m+1} = l']\, P(J_n = k | J_{m+1} = l') \\ &= \sum_{m=0}^{n-1} \sum_{l,l' \in E} P^{(m)}_{jl} E_l[X_1; J_1 = l'] P^{(n-1-m)}_{l'k}. \end{aligned}$$

□

The following is analogous to the Strong Law of Large Numbers for the additive component of the MRW. More limit theorems which are analogous to those for random walks will be presented in Chap. 5.

Theorem 4.10. *Assume that the condition (4.36) holds and the stationary probability distribution $\{\pi_k, k \in E\}$ exists. Then S_n/n converges a.s., and also in $\mathcal{L}^1$. Thus*

$$\lim_{n\to\infty} \frac{S_n}{n} = \sum_{k \in E} \pi_k E_k(X_1) \quad a.s. \tag{4.42}$$

and

$$\lim_{n\to\infty} E_j\left[\frac{S_n}{n}\right] = \sum_{k \in E} \pi_k E_k(X_1). \tag{4.43}$$

Proof. Since for each fixed $j \in E$, the $\{\tau_r^j\}$ are stopping times for $\{(S_n, J_n)\}$ they form a convenient choice of an imbedding renewal process. Therefore $\{S_n\}$ is a regenerative process. Also since the stationary distribution of J exists, $E_j(\tau_1^j) = \pi_j^{-1} < \infty$, for all j. In view of (4.36) we find from Lemma 4.6 that $E_j(|S_1^j|) < \infty$. Using the regenerative property of the $\{S_n\}$ we find that

$$\lim_{n\to\infty} \frac{S_n}{n} = \frac{E(\text{value in a cycle})}{E(\text{cycle length})} = \frac{E_j(S_1^j)}{E_j(\tau_1^j)} = \pi_j E_j(S_1^j)$$

and (4.42) follows from (4.38).

Moreover, we conclude (4.43) from the fact that Lemma 4.7 gives

$$E_j[S_n] = \sum_{m=0}^{n-1} \sum_{l\in E} P_{jl}^{(m)} E_l[X_1]$$

and, using (4.36) and applying the dominated convergence theorem, we find that

$$\lim_{n\to\infty} E_j\left[\frac{S_n}{n}\right] = \sum_{l\in E} E_l[X_1] \lim_{n\to\infty} \frac{1}{n} \sum_{m=0}^{n-1} P_{jl}^{(m)} = \sum_{l\in E} \pi_l E_l(X_1). \qquad \square$$

The above result yields the following.

Theorem 4.11. *Suppose that the stationary distribution of J exists and (4.36) holds. Then $\{S_n\}$ either*

(i) *Drifts to $+\infty$;*
(ii) *Drifts to $-\infty$; or*
(iii) *Oscillates in the sense of (4.14)*

according as $\sum_{k\in E} \pi_k E_k(X_1)$ is positive, negative, or zero.

Proof. The results (i) and (ii) follow directly from Theorem 4.10, while (iii) corresponds to the case where the limit is zero which, from (4.38), is always the case for degenerate MRW. Hence, by Theorem 4.3 the result follows for degenerate MRW. In the case of nondegenerate MRW, at least one of the imbedded random walks, say $\{S_r^j\}$, is nondegenerate and the result follows from the fact that

$$\liminf_n S_n \le \liminf_r S_r^j = -\infty \quad \text{and} \quad \limsup_n S_n \ge \limsup_r S_r^j = +\infty. \qquad \square$$

Remark 4.6. It can be seen that the above result is almost a complete analogue to that of the standard random walk except that in the case of degenerate MRW, $\liminf_n S_n$ and $\limsup_n S_n$ may be finite.

4.7 Renewal Equation

In this section, we consider a renewal equation for the MRW as a natural generalization of that for a MRP. The MRW integral equation is defined as

$$Z_j(A) = f_j(A) + \sum_{k \in E} \int_{\mathbf{R}} Q_{jk}\{dy\} Z_k(A-y) \quad (j \in E) \tag{4.44}$$

where $f_j(A)$ are positive bounded functions over finite intervals for each $A \in \mathcal{B}$, where $(\mathbf{R}, \mathcal{B})$ denotes the Borel σ-field on the reals, and $j \in E$. In vector form we write

$$Z(A) = f(A) + Q * Z(A). \tag{4.45}$$

We seek to establish the necessary and sufficient conditions for its unique solution to exist. First we need the following lemma. For each fixed $A \in \mathcal{B}$, let

$$q_j^{(n)}(A) = P_j\{S_n \in A\} = \sum_{k \in E} Q_{jk}^{(n)}(A) \tag{4.46}$$

$$q_j(A) = \lim_{n \to \infty} q_j^{(n)}(A). \tag{4.47}$$

Then we have the following result, where we let

$$\mathcal{B}_0 = \{A \in \mathcal{B} : 0 \in A \text{ and } A \text{ is bounded}\}.$$

Lemma 4.8. $q = \{q_j(A),\ j \in E, A \in \mathcal{B}_0\}$ *is the maximal solution of the integral equation*

$$x_j(A) = \sum_{k \in E} \int_{\mathbf{R}} Q_{jk}\{dy\} x_k(A-y) \tag{4.48}$$

with $0 \le x_j(A) \le 1$, *for all* $j \in E$ *and* $A \in \mathcal{B}_0$.

Proof. The regularity condition on the transition measure implies that

$$q_j^{(0)}(A) = \mathbf{1}_{\{0 \in A\}} = 1$$

for all $A \in \mathcal{B}_0$, and it follows from the Chapman-Kolmogorov equation that

$$q_j^{(n+1)}(A) = \sum_{k \in E} \int_{\mathbf{R}} Q_{jk}\{dy\} q_k^{(n)}(A-y) \quad (n \ge 0).$$

By the bounded convergence theorem, $q_j(A)$ satisfies the above integral equation. For any solution to (4.48) with $0 \le x_j(A) \le 1$, we have

$$x_j(A) \le q_j^{(0)}(A) = 1 \quad (j \in E, A \in \mathcal{B}_0)$$

which in turn implies that

$$x_j(A) \leq \sum_{k\in E} \int_A Q_{jk}\{dy\} = q_j^{(1)}(A) \quad (j \in E, A \in \mathcal{B}_0).$$

So inductively, $x_j(A) \leq q_j^{(n)}(A)$ by repeated substitution. Taking limit for $n \to \infty$, we arrive at the desired result. □

Next, for each $A \in \mathcal{B}$ and $j, k \in E$, we define the following counting processes

$$N_k(A) = \sum_{n=1}^{\infty} \mathbf{1}_{\{S_n \in A, J_n = k\}} = \#\{n > 0 : (S_n, J_n) \in A \times \{k\}\}$$

$$N(A) = \sum_{k\in E} N_k(A) = \#\{n > 0 : S_n \in A\}$$

the corresponding renewal measures

$$U_{jk}(A) = E_j[N_k(A)] = \sum_{n=1}^{\infty} Q_{jk}^{(n)}(A)$$

$$U_j(A) = E_j[N(A)] = \sum_{n=1}^{\infty} q_j^{(n)}(A)$$

and let $U(A) = (U_{jk}(A))_{E\times E}$.

Theorem 4.12. *A particular solution of (4.44) is given by*

$$G = U * f$$

and the general solution is of the form

$$Z = G + h$$

where h satisfies the equation

$$h = Q * h.$$

Proof. Since (4.44) is an implicit function of Z, by repeated substitution into (4.44) for Z on the right hand side, we have

$$Z = \sum_{r=0}^{n} Q^r * f + Q^{n+1} * Z.$$

As $n \to \infty$, by monotone convergence theorem the right hand side becomes

$$U * f + h$$

which is the desired result. □

Corollary 4.2. *The solution of MRW equation is unique iff* $q(A) = 0$ *for all* $A \in \mathcal{B}_0$.

Proof. It follows from Lemma 4.8 that $q(A)$ is the maximal solution of (4.48) and $h(A)$ satisfies (4.48). In this case the unique solution is given by $G = U * f$. □

The natural question is then when does $q(A) = 0$ for all $A \in \mathcal{B}_0$. The answer is in Sec. 4.5 on fluctuation theory. It can be seen that $q(A) \equiv 0$ iff S_n does not oscillate. This gives us the following.

Theorem 4.13. *Suppose that the MRW is nondegenerate and the stationary distribution of* J *exists.*

Then the solution of MRW equation is unique iff

(i) Any of the imbedded random walks is drifting;
(ii) Only one of the semirecurrent sets, ζ_- *or* ζ_+, *is nonterminating.*

Proof. It follows from theorems 4.8 and 4.11 that these are the 2 cases that S_n does not oscillate. □

4.8 Wiener-Hopf Factorization

Next we obtain a factorization in terms of measures by considering the semirecurrent sets similar to those defined in (4.23). To begin with, we derive the following result which is a direct consequence of Lemma 4.3.

Lemma 4.9. *For every finite interval* I *and* $j, k \in E$, *we have*

$$\begin{aligned} &P_k\{S_m \leq 0 \ (0 \leq m \leq n), S_n \in I, J_n = j\} \\ &\quad = \frac{\pi_j}{\pi_k} P_j\{\widehat{S}_n - \widehat{S}_{n-m} \leq 0 \ (0 \leq m \leq n), \widehat{S}_n \in I, \widehat{J}_n = k\}. \end{aligned} \tag{4.49}$$

Proof. Let

$$\mathcal{A}_n = \{\omega : S_1 \leq 0, S_2 \leq 0, \ldots, S_n \leq 0, S_n \in I\}.$$

For the time-reversed MRW $(\widehat{S}_n, \widehat{J}_n)$ we have

$$S_m = \widehat{S}_n - \widehat{S}_{n-m} \ (0 \leq m \leq n); \quad S_n = \widehat{S}_n; \quad \widehat{J}_n = k; \quad \widehat{J}_0 = j$$

so the set $\widehat{\mathcal{A}}_n$ corresponding to $\mathcal{A}_n$ as in Lemma 4.3 is given by

$$\widehat{\mathcal{A}}_n = \{\omega : \widehat{S}_n - \widehat{S}_{n-m} \leq 0 \ \ (0 \leq m \leq n), \widehat{S}_n \in I\}.$$

The desired result follows from Lemma 4.3. □

With a slight change in notation, let $\zeta^+(\widehat{\zeta}^+)$ denote the strong ascending ladder set of the MRW (time-reversed MRW); thus

$$\zeta^+ = \{(n,j) : S_n = M_n, J_n = j, \rho_n = n\} \tag{4.50}$$

where ρ_n is the first epoch at which $\{S_m\}$ attains M_n.

For fixed $s \in (0,1)$, we define the following matrix-measures μ_s^+, $\widehat{\mu}_s^-$, and μ_s, where $\mu_s^+\{I\} = (\mu_{jk}^+\{I\})_{E\times E}$, $\widehat{\mu}_s^-\{I\} = (\widehat{\mu}_{jk}^-\{I\})_{E\times E}$ and $\mu_s\{I\} = (\mu_{jk}\{I\})_{E\times E}$, for a finite interval I, as follows:

$$\mu_{jk}^+\{I\} = \sum_{n=0}^{\infty} E_j(s^n; (n,k) \in \zeta^+, S_n \in I, J_n = k) \tag{4.51}$$

$$\widehat{\mu}_{jk}^-\{I\} = \sum_{n=0}^{\infty} E_j(s^n; (n,k) \in \widehat{\zeta}^-, \widehat{S}_n \in I, \widehat{J}_n = k) \tag{4.52}$$

$$\mu_{jk}\{I\} = \sum_{n=0}^{\infty} s^n Q_{jk}^{(n)}\{I\}. \tag{4.53}$$

Note that each of these measures is finite, a simple bound being $(\mathbf{I} - s\mathbf{P})^{-1}$, where $\mathbf{P}$ is the transition probability matrix of J. We then have the following result.

Theorem 4.14. *We have*

$$\mu_s = \mu_s^+ * \nu_s^- \tag{4.54}$$

where

$$\nu_s^-\{I\} = \left\{(\nu_{jk}^-\{I\})_{E\times E} : \nu_{jk}^-\{I\} = \frac{\pi_k}{\pi_j}\widehat{\mu}_{kj}^-\{I\}\right\}. \tag{4.55}$$

In particular,

$$\mathbf{U}(s)\mathbf{V}(s) = (\mathbf{I} - s\mathbf{P})^{-1} \tag{4.56}$$

with

$$\mathbf{U}(s) = \mu_s^+\{(0,\infty)\} \quad \textit{and} \quad \mathbf{V}(s) = \nu_s^-\{(-\infty, 0]\}. \tag{4.57}$$

Proof. First we note that

$$\begin{aligned} Q_{ij}^{(n)}\{I\} &= \sum_{k\in E}\sum_{l=0}^{n} P_i\{\rho_n = l, J_{\rho_n} = k, S_n \in I, J_n = j\} \\ &= \sum_{k\in E}\sum_{l=0}^{n} \int_0^{\infty} P_i\{S_m < S_l \ \ (0 \le m \le l-1), J_l = k, S_l \in dx\} \\ &\quad \cdot P_i\{S_m \le S_l \ \ (l \le m \le n), J_n = j, S_n \in I | S_l = x, J_l = k\} \end{aligned}$$

The second term in the last expression turns out to be

$$P_k\{S_m \leq 0 \ \ (0 \leq m \leq n-l), J_{n-l} = j, S_{n-l} \in I - x\}$$

and by Lemma 4.5 this equals

$$\begin{aligned}\frac{\pi_j}{\pi_k} P_j\{\widehat{S}_{n-l} - \widehat{S}_{n-l-m} \leq 0 \ \ (0 \leq m \leq n-l), \widehat{S}_{n-l} \in I - x, \widehat{J}_{n-l} = k\} \\ = \frac{\pi_j}{\pi_k} P_j\{(n-l, k) \in \widehat{\zeta}^-, \widehat{S}_{n-l} \in I - x, \widehat{J}_{n-l} = k\}.\end{aligned}$$

It follows that

$$\begin{aligned}\mu_{ij}\{I\} &= \sum_{n=0}^{\infty} s^n Q_{ij}^{(n)}\{I\} \\ &= \sum_{n=0}^{\infty} s^n \sum_{k \in E} \sum_{l=0}^{n} \int_0^{\infty} P_i\{(l,k) \in \zeta^+, J_l = k, S_l \in dx\} \\ &\qquad \cdot \frac{\pi_j}{\pi_k} P_j\{(n-l, k) \in \widehat{\zeta}^-, \widehat{S}_{n-l} \in I - x, \widehat{J}_{n-l} = k\} \\ &= \sum_{k \in E} \int_0^{\infty} \mu_{ik}^+\{dx\} \frac{\pi_j}{\pi_k} \widehat{\mu}_{jk}^-\{I - x\} \\ &= \sum_{k \in E} \int_0^{\infty} \mu_{ik}^+\{dx\} \nu_{kj}^-\{I - x\} \\ &= (\mu_s^+ * \nu_s^-)_{ij}\{I\}.\end{aligned}$$

This establishes the main factorization (4.54). As a special case we have

$$\mu_s\{\mathbf{R}\} = \mu_s^+\{\mathbf{R}_+\}\nu_s^-\{\mathbf{R}_-\} = \mathbf{U}(s)\mathbf{V}(s).$$

Thus (4.56) follows as, on the other hand,

$$\begin{aligned}\mu_s\{\mathbf{R}\} &= \sum_{n=0}^{\infty} s^n \mathbf{Q}^{(n)}\{\mathbf{R}\} \\ &= \sum_{n=0}^{\infty} s^n \mathbf{P}^n = (\mathbf{I} - s\mathbf{P})^{-1}.\end{aligned}$$

□

As a direct consequence of Theorem 4.14 we arrive at a Wiener-Hopf factorization obtained in terms of transforms by Presman [97] who called it the basic factorization identity. Let

$$\chi^+(s,\omega) = \{(\chi_{jk}^+(s,\omega))_{E\times E} : \chi_{jk}^+(s,\omega) = E_j(s^T e^{i\omega S_T}; J_T = k)\} \quad (4.58)$$

$$\chi^-(s,\omega) = \{(\chi_{jk}^-(s,\omega))_{E\times E} : \chi_{jk}^-(s,\omega) = E_j(s^{\bar{T}} e^{i\omega S_{\bar{T}}}; J_{\bar{T}} = k)\} \quad (4.59)$$

where $T = T_1$ and $\bar{T} = \bar{T}_1$ and, as usual, $\widehat{\chi}^+(s,\omega)$ and $\widehat{\chi}^-(s,\omega)$ denote the counterparts for the corresponding time-reversed MRW.

The following result is a consequence of Theorem 4.5 stating that $\{(S_{T_k}, T_k, J_{T_k})\}$ and $\{(-S_{\bar{T}_k}, \bar{T}_k, J_{\bar{T}_k})\}$ are MRPs.

Lemma 4.10. *For $r \geq 1$, we have*

$$E_j(s^{T_r} e^{i\omega S_{T_r}}; J_{T_r} = k) = \left[(\chi^+(s,\omega))^r\right]_{jk} \tag{4.60}$$

$$E_j(s^{\bar{T}_r} e^{i\omega S_{\bar{T}_r}}; J_{\bar{T}_r} = k) = \left[(\chi^-(s,\omega))^r\right]_{jk}. \tag{4.61}$$

The above lemma provides us enough machinery to write down a factorization in terms of transforms. Define

$$\Phi(\omega) = \left\{(\phi_{jk}(\omega))_{E\times E} : \phi_{jk}(\omega) = \int_{-\infty}^{\infty} e^{i\omega x} Q_{jk}\{dx\}\right\}.$$

We have then the following.

Theorem 4.15. *For the MRW with J having a stationary probability distribution $\{\pi_i\}$ we have*

$$\mathbf{I} - s\Phi(\omega) = \mathbf{D}_\pi^{-1}[\mathbf{I} - \widehat{\chi}^-(s,\omega)]'\mathbf{D}_\pi[\mathbf{I} - \chi^+(s,\omega)] \tag{4.62}$$

where $\mathbf{D}_\pi = diag(\pi)$ and the prime denotes the transpose.

Proof. We have

$$\sum_{n=0}^{\infty} \int_{-\infty}^{\infty} e^{i\omega x} s^n \mathbf{Q}^{(n)}\{dx\} = \sum_{n=0}^{\infty} s^n [\Phi(\omega)]^n = [\mathbf{I} - s\Phi(\omega)]^{-1}.$$

From Lemma 4.10, the transform of μ_{jk}^+ is

$$\int_{-\infty}^{\infty} e^{i\omega x} \mu_{jk}^+\{dx\} = \sum_{r=0}^{\infty} E_j(s^{T_r} e^{i\omega S_{T_r}}; J_{T_r} = k) = \sum_{r=0}^{\infty} \left[(\chi^+(s,\omega))^r\right]_{jk}$$

so that

$$\int_{-\infty}^{\infty} e^{i\omega x} \mu_s^+\{dx\} = [\mathbf{I} - \chi^+(s,\omega)]^{-1}.$$

Similarly the transform of $\widehat{\mu}_s^-$ is $[\mathbf{I} - \widehat{\chi}^-(s,\omega)]^{-1}$. From (4.55), an easy manipulation gives the transform of ν_s^- as $[\mathbf{D}_\pi^{-1}[\mathbf{I} - \widehat{\chi}^-(s,\omega)]'\mathbf{D}_\pi]^{-1}$.

Starting from (4.54), the desired result then follows by trivial algebraic rearrangement. □

4.9 First Exit Time From a Bounded Interval

We have seen that the fluctuation behavior of a non-degenerate MRW with J persistent is similar to that of a standard random walk. It is then interesting to investigate the corresponding results for the first exit time from a finite interval. Let $I = (-a, b)$ with $a, b > 0$ and

$$N = \min\{n : S_n \notin I\}$$

so that N is the first time the additive component of the MRW exits the interval I. In the case of a standard random walk, it is known that:

(i) $P\{N > n\}$ is geometrically bounded and hence $P\{N = \infty\} = 0$.
(ii) The moment generating function of N exists.

We derive the analogous results for the MRW.

Lemma 4.11. *Suppose that the MRW is nondegenerate with J persistent. Then there exists a decreasing sequence $\{B_n\}_{n>0}$ taking values on the interval $[0,1]$ such that*

$$P\{N > n\} \leq B_n \qquad (n > 0) \tag{4.63}$$

regardless of J_0.

Proof. For $j \in E$ and $n > 0$,

$$\begin{aligned} P_j\{N > n\} &= P_j\{S_1 \in I, S_2 \in I, \ldots, S_n \in I\} \\ &\leq P_j\{S_r^j \in I, r = 1, 2, \ldots, N_n^j\} \end{aligned} \tag{4.64}$$

where N_n^j is the number of visits to j from time 1 up to time n. Since S_r^j is an nonterminating standard random walk for each j, we have by Stein's lemma

$$E_j[\mathbf{1}_{\{N>n\}} | N_n^j = m_j] \leq A_j \delta_j^{m_j+1} \tag{4.65}$$

where $A_j > 0$ and $0 < \delta_j < 1$. Taking expectation on both sides of (4.65) we find that

$$P_j\{N > n\} \leq A_j E\left[\delta_j^{N_n^j+1}\right]. \tag{4.66}$$

Since N_n^j increases with n, it is clear that $E[\delta_j^{N_n^j+1}]$ decreases with n.

Now let $B_n = \inf_j A_j E[\delta_j^{N_n^j+1}]$. The desired result follows from the fact that $B_{n+1} \leq B_n$, for $n \geq 1$. $\square$

Theorem 4.16. *Suppose that the MRW is nondegenerate with J persistent. Then we have*

$$N < \infty \quad a.s.$$

If in addition the stationary distribution of J exists, then

$$E_j[N^\alpha] < \infty \qquad (\alpha > 0, j \in E)$$

i.e., N has a proper distribution with finite moments of all orders.

Proof. To prove that $N < \infty$ a.s., we only need to show that the sequence $\{B_n\}_{n>0}$ in the proof of Lemma 4.11 tends to zero as n tends to infinity. This result follows from the fact that since J is persistent, $N_n^j \to \infty$ as $n \to \infty$, for all $j \in E$. Taking into account (4.66), we have

$$\begin{aligned} E_j[N^\alpha] &= \sum_{n=1}^{\infty} n^\alpha P_j\{N = n\} \\ &\leq \sum_{n=1}^{\infty} n^\alpha P_j\{N \geq n\} \\ &\leq \sum_{n=1}^{\infty} n^\alpha A_j E_j[\delta_j^{N_n^j}] \\ &= A_j \sum_{n=1}^{\infty} E_j\left[\left(e^{\alpha \frac{\log n}{n}} \delta_j^{\frac{N_n^j}{n}}\right)^n\right]. \end{aligned} \tag{4.67}$$

Suppose now that the stationary distribution of J exists. Then, as

$$\frac{\log n}{n} \underset{n\to\infty}{\longrightarrow} 0 \quad \text{and} \quad \frac{N_n^j}{n} \underset{n\to\infty}{\longrightarrow} \pi_j \ a.s.$$

it follows that

$$e^{\alpha \frac{\log n}{n}} \delta_j^{\frac{N_n^j}{n}} \underset{n\to\infty}{\longrightarrow} \delta_j^{\pi_j} \quad a.s.$$

This implies that given $0 < \epsilon < 1 - \delta_j^{\pi_j}$, theres exists m_j such that for all $n > m_j$,

$$e^{\alpha \frac{\log n}{n}} \delta_j^{\frac{N_n^j}{n}} < \delta_j^{\pi_j} + \epsilon$$

thus implying that for all $n > m_j$

$$E_j\left[\left(e^{\alpha \frac{\log n}{n}} \delta_j^{\frac{N_n^j}{n}}\right)^n\right] < \left(\delta_j^{\pi_j} + \epsilon\right)^n.$$

Thus, as $\delta_j^{\pi_j} + \epsilon < 1$, we conclude, in view of (4.67), that $E_j[N^\alpha] < \infty$, for all $j \in E$. □

Chapter 5

Limit Theorems for Markov Random Walks

In this chapter we obtain the central limit theorem and law of iterated logarithm for the additive component of a Markov random walk. The parameters involved are computed explicitly in terms of the transition measure of the Markov random walk. An application in finance is briefly discussed.

5.1 Introduction

Suppose that we are given a probability space $(\Omega, \mathcal{F}, P)$. We denote $\mathbf{R} = (-\infty, \infty)$ and E = countable set. A time-homogeneous Markov random walk (MRW) $(S, J) = \{(S_n, J_n), n \geq 0\}$ is a time-homogeneous Markov process on the state space $\mathbf{R} \times E$ whose transition distribution measure is given by

$$P\{(S_{m+n}, J_{m+n}) \in A \times \{k\} | (S_m, J_m) = (x, j)\}$$
$$= P\{(S_n, J_n) \in (A - x) \times \{k\} | J_0 = j\} \qquad (5.1)$$

for all $j, k \in E$ and a Borel subset A of $\mathbf{R}$. We denote this probability as $Q_{jk}^{(n)}\{A - x\}$ and assume that it satisfies the regularity conditions. Also we shall denote the conditional probabilities and expectations given $J_0 = j$ as P_j and E_j respectively. Henceforth, we assume that the stationary distribution of J, $\{\pi_k, k \in E\}$, exists and

$$\sum_{k \in E} \pi_k E_k[|X_1|] < \infty \qquad (5.2)$$

where $X_n = S_n - S_{n-1}$ denotes the nth increment of the MRW.

[5] This chapter is based on the article by L.C. Tang, 'Limit theorems for Markov random walks', *Statist. Probab. Lett.* **18**, 4, pp. 265–270, 1993, and contents are reproduced with permission of the publisher, Elsevier.

Keilson and Wishart [52] obtained a central limit theorem for the MRW using matrix analysis where the result holds only when the modulating Markov chain J is on finite state space and is ergodic. They expressed the parameters involved as functions of eigenvalues derived from the transition measure of the MRW, which are difficult to compute algebraically. Independent from their work, Anisimov [3] also obtained some results on limit theorems for MRWs on a modulating Markov chain with finite state space. More recently, Prabhu, Tang and Zhu [95] presented some new results for the MRW (see Chap. 4). Among them is the strong law of large numbers for the additive component S_n of the MRW.

In this chapter, we establish the central limit theorem and law of iterated logarithm for the additive component of an MRW. Our proofs are probabilistic using i.i.d. sequences of random variables obtained from the MRW. This idea is analogous to that of additive functionals of Markov chains discussed in Chung [26]. We express the parameters involved in terms of the transition measure of the MRW. Finally, we indicate an application of these theorems to finance models.

5.2 A Sequence of Normalized Increments

To obtain the central limit theorem for the MRW, we first construct the following i.i.d. sequence of random variables denoting normalized increments of the additive component of the MRW between successive visits of the modulating Markov chain to a given state. For each $j \in E$, let

$$Y_r^j = (S_{r+1}^j - S_r^j) - \mu(\tau_{r+1}^j - \tau_r^j) \qquad (r \geq 1) \tag{5.3}$$

where $\tau_0^j = 0$ and

$$\tau_r^j = \min\{n > \tau_{r-1}^j : (S_n, J_n) \in \{\mathbf{R}\} \times \{j\}\} \qquad (r \geq 1)$$

are the successive hitting times of state j, $S_r^j = S_{\tau_r^j}$ $(r \geq 0)$, and $\mu = \pi_j E_j(S_1^j)$ denotes the expected increment in the additive component of the MRW per time unit. Note that, as in (4.38), we have

$$\mu = \pi_j E_j(S_1^j) = \sum_{k \in E} \pi_k E_k(X_1). \tag{5.4}$$

We also denote the number of hits (visits) to state j from time 1 up to time n as

$$N_n^j = \max\{r : \tau_r^j \leq n\}. \tag{5.5}$$

For the sequence $\{Y_r^j, r \geq 1\}$ defined above, we have the following.

Lemma 5.1. *The random variables of the sequence $\{Y_r^j, r \geq 1\}$ are i.i.d. with mean zero and variance given by*

$$\begin{aligned} Var(Y_1^j) = & \sum_{k\in E} {}^jP_{jk}^* E_k(X_1^2) \\ & + 2\sum_{i,k\neq j} {}^jP_{ji}^* E_i(X_1; J_1 = k) \sum_{l\in E} {}^jP_{kl}^* E_l(X_1) \\ & + \mu^2 \Big[\sum_{k\in E} {}^jP_{jk}^* + 2\sum_{k\neq j} {}^jP_{jk}^* \sum_{l\in E} {}^jP_{kl}^*\Big] \\ & - 2\mu \sum_{k\in E} E_k(X_1)[\,{}^jP_{jk}^* + {}^j\tilde{P}'_{jk}(1)] \end{aligned} \tag{5.6}$$

where

$${}^jP_{lk}^* = \sum_{n=0}^{\infty} {}^jP_{lk}^{(n)}, \qquad {}^j\tilde{P}_{jk}(s) = \sum_{n=0}^{\infty} {}^jP_{jk}^{(n)} s^n, \qquad {}^j\tilde{P}'_{jk}(1) = \lim_{s\to 1} \frac{d}{ds}[\,{}^j\tilde{P}_{jk}(s)]$$

with

$${}^jP_{lk}^{(n)} = P_l(J_n = k; \tau_1^j > n)$$

provided that all the sums converge.

Proof. (i). We note that the sequence $\{\tau_r^j, r \geq 1\}$ is a renewal process for each j and $\{S_r^j\}$ is a random walk on the state space $\mathbf{R}$, so that $\{Y_r^j\}$ is an i.i.d. sequence of random variables. Taking expectation on the right side of (5.3) we have

$$E(S_{r+1}^j - S_r^j) - \pi_j E_j(S_1^j) E(\tau_{r+1}^j - \tau_r^j) = E_j(S_1^j) - \pi_j E_j(S_1^j) \cdot \frac{1}{\pi_j} = 0.$$

(ii). We have

$$\mathrm{Var}(Y_1^j) = E_j[(Y_1^j)^2] = E_j[(S_1^j)^2 + \mu^2(\tau_1^j)^2 - 2\mu S_1^j \tau_1^j] \tag{5.7}$$

since $Y_1 \overset{\mathrm{d}}{=} S_1^j - \mu\tau_1^j$ on $\{J_0 = j\}$, where $\overset{\mathrm{d}}{=}$ denotes equality in distribution. As

$$\begin{aligned} E_j[(S_1^j)^2] &= E_j\left[\left(\sum_{m=1}^{\infty} X_m \mathbf{1}_{\{\tau_1^j \geq m\}}\right)^2\right] \\ &= E_j\Big[\sum_{m=1}^{\infty} X_m^2 \mathbf{1}_{\{\tau_1^j \geq m\}} + 2\sum_{m=1}^{\infty} X_m \sum_{r=1}^{\infty} X_{m+r} \mathbf{1}_{\{\tau_1^j \geq m+r\}}\Big] \\ &= \sum_{m=1}^{\infty} E_j(X_m^2 \mathbf{1}_{\{\tau_1^j \geq m\}}) \\ &\quad + 2E_j\Big[\sum_{m=1}^{\infty} \sum_{i,k\in E} X_m \mathbf{1}_{\{J_{m-1}=i, J_m=k\}} \sum_{r=1}^{\infty} X_{m+r} \mathbf{1}_{\{\tau_1^j \geq m+r\}}\Big] \end{aligned}$$

$$= \sum_{m=1}^{\infty} E_j(X_m^2 \mathbf{1}_{\{\tau_1^j \geq m\}})$$

$$+ 2 \sum_{m=1}^{\infty} \sum_{i,k \neq j} {}^jP_{ji}^{(m-1)} E_i(X_1; J_1 = k) \sum_{r=1}^{\infty} E_k(X_r \mathbf{1}_{\{\tau_1^j \geq r\}}) \tag{5.8}$$

$$= \sum_{k \in E} {}^jP_{jk}^* E_k(X_1^2)$$

$$+ 2 \sum_{i,k \neq j} \sum_{m=1}^{\infty} {}^jP_{ji}^{(m-1)} E_i(X_1; J_1 = k) \sum_{l \in E} \sum_{r=1}^{\infty} {}^jP_{kl}^{(r-1)} E_l(X_1)$$

we then have

$$E_j\left[(S_1^j)^2\right] = \sum_{k \in E} {}^jP_{jk}^* E_k(X_1^2)$$

$$+ 2 \sum_{i,k \neq j} {}^jP_{ji}^* E_i(X_1; J_1 = k) \sum_{l \in E} {}^jP_{kl}^* E_l(X_1). \tag{5.9}$$

Substituting 1 for $E_k(X_1)$ and $E_k(X_1^2)$ and proceeding from (5.8), we have

$$E_j\left[(\tau_1^j)^2\right] = \sum_{k \in E} {}^jP_{jk}^* + 2 \sum_{k \neq j} {}^jP_{jk}^* \sum_{l \in E} {}^jP_{kl}^*. \tag{5.10}$$

Finally, we have

$$E_j(S_1^j \tau_1^j) = E_j\left[\sum_{n=1}^{\infty} n \mathbf{1}_{\{\tau_1^j = n\}} \sum_{m=1}^{n} X_m\right]$$

$$= \sum_{n=1}^{\infty} n \sum_{m=1}^{n} E_j(X_m; \tau_1^j = n)$$

$$= \sum_{n=1}^{\infty} \sum_{r=1}^{n} \sum_{m=1}^{n} E_j(X_m; \tau_1^j = n)$$

$$= \sum_{r=1}^{\infty} \sum_{n=r}^{\infty} \sum_{m=1}^{n} E_j(X_m; \tau_1^j = n)$$

$$= \sum_{r=1}^{\infty} \sum_{m=r}^{\infty} \sum_{n=m}^{\infty} E_j(X_m; \tau_1^j = n)$$

$$= \sum_{r=1}^{\infty} \sum_{m=r}^{\infty} E_j(X_m; \tau_1^j \geq m)$$

$$= \sum_{r=1}^{\infty} \sum_{m=r}^{\infty} \sum_{k \in E} E_j(X_m; \tau_1^j > m-1; J_{m-1} = k)$$

$$= \sum_{k \in E} E_k(X_1) \sum_{r=1}^{\infty} \sum_{m=r}^{\infty} P_j\{\tau_1^j > m-1; J_{m-1} = k\}$$

$$= \sum_{k \in E} E_k(X_1) \sum_{m=1}^{\infty} \sum_{r=1}^{m} P_j\{\tau_1^j > m-1; J_{m-1} = k\}$$

$$= \sum_{k \in E} E_k(X_1) \sum_{m=0}^{\infty} (m+1) P_j\{\tau_1^j > m; J_m = k\}$$

$$= \sum_{k \in E} E_k(X_1) \left[{}^jP_{jk}^* + \sum_{m=0}^{\infty} m P_j\{\tau_1^j > m; J_m = k\} \right]. \tag{5.11}$$

Therefore

$$E_j(S_1^j \tau_1^j) = \sum_{k \in E} E_k(X_1)[{}^jP_{jk}^* + {}^j\tilde{P}'_{jk}(1)]. \tag{5.12}$$

The result follows by substituting (5.9), (5.10), and (5.12) into (5.7). □

Remark 5.1. Using the fact that

$$E_j(\tau_1^j) = \frac{1}{\pi_j}, \qquad {}^jP_{jk}^* = \frac{\pi_k}{\pi_j}, \qquad \sum_{l \in E} {}^jP_{kl}^* = E_k(\tau_1^j)$$

and

$${}^jP_{kl}^* = \pi_l(E_k(\tau_1^j) + E_j(\tau_1^l) - E_k(\tau_1^l))$$

we may express (5.9) and (5.10) in terms of the stationary distribution of J as follows.

$$E_j[(S_1^j)^2] = \frac{1}{\pi_j} \Bigg\{ \sum_{k \in E} \pi_k E_k(X_1^2) + 2 \sum_{i,k \neq j} \pi_i E_i(X_1; J_1 = k)$$

$$\cdot \left[\sum_{l \in E} E_l(X_1) \pi_l \left[E_k(\tau_1^j) + E_j(\tau_1^l) - E_k(\tau_1^l) \right] \right] \Bigg\}. \tag{5.13}$$

$$E_j(\tau_1^j)^2 = \frac{1}{\pi_j} \left[1 + 2 \sum_{k \neq j} \pi_k E_k(\tau_1^j) \right]. \tag{5.14}$$

5.3 Limit Theorems

In this section we establish the central limit theorem and law of iterated logarithm for the additive component of the MRW.

Theorem 5.1. (Central Limit Theorem for MRW)
If $\sigma^2 = \pi_j \, Var(Y_1^j) \in (0, \infty)$, then

$$\frac{S_n - n\mu}{\sigma\sqrt{n}} \xrightarrow{\mathcal{L}} \mathcal{N}$$

where $\mathcal{N}$ denotes the standard normal distribution and $\xrightarrow{\mathcal{L}}$ denotes convergence in distribution.

Proof. We write

$$S_n - n\mu = \sum_{m=1}^{\tau_1^j} X_m + \sum_{r=1}^{N_n^j - 1} Y_r^j + \sum_{m=\tau_{N_n^j}^j + 1}^{n} X_m + \mu(\tau_{N_n^j}^j - \tau_1^j - n). \quad (5.15)$$

Dividing it by $\sigma\sqrt{n}$ and let $n \to \infty$ we have the 1st term tends to zero as it does not dependent on n. For the 3rd term we note that, proceeding as in the proof of Lemma 5.1 and using the fact that ${}^jP_{jk}^* = \pi_k/\pi_j$,

$$E\left[\left|\sum_{m=\tau_{N_n^j}^j+1}^{n} X_m\right|\right] \le E\left[\sum_{m=\tau_{N_n^j}^j+1}^{n} |X_m|\right]$$
$$\le E_j\left[\sum_{m=1}^{\tau_1^j} |X_m|\right] = \sum_{k\in E} \frac{\pi_k}{\pi_j} E_k[|X_1|] < \infty$$

in view of (5.2). It then follows that $\sum_{m=\tau_{N_n^j}^j+1}^{n} X_m$ is bounded in probability and thus $\sum_{m=\tau_{N_n^j}^j+1}^{n} X_m/n$ converges to 0. For the last term, it also converges to 0 since

$$\frac{\tau_{N_n^j} - \tau_1^j}{n} = \frac{N_n^j}{n} \cdot \frac{\sum_{r=2}^{N_n^j}(\tau_r^j - \tau_{r-1}^j)}{N_n^j - 1} \cdot \frac{N_n^j - 1}{N_n^j} \to \pi_j \cdot \frac{1}{\pi_j} \cdot 1 = 1.$$

Finally, we have the 2nd term converges in distribution to the standard normal distribution by the central limit theorem with random index; i.e.

$$\sqrt{\frac{N_n^j - 1}{N_n^j}} \sqrt{\frac{N_n^j}{n\pi_j}} \frac{\sum_{r=1}^{N_n^j - 1} Y_r^j}{\sqrt{(N_n^j - 1)\mathrm{Var}(Y_1^j)}} \xrightarrow{\mathcal{L}} \mathcal{N}$$

since $(N_n^j - 1)/N_n^j \to 1$ and $N_n^j/(n\pi_j) \to 1$. □

Theorem 5.2. (Law of Iterated Logarithm for MRW)
Under the conditions of Theorem 5.1, we have

$$\limsup \frac{S_n - n\mu}{\sqrt{2\sigma^2 n \log\log n}} = 1 \quad a.s. \tag{5.16}$$

and

$$\liminf \frac{S_n - n\mu}{\sqrt{2\sigma^2 n \log\log n}} = -1 \quad a.s. \tag{5.17}$$

Proof. Applying the law of iterated logarithm for random walks on the sequence $\{Y_r^j\}$, we have

$$\limsup \frac{S_n - n\mu}{\sqrt{2\sigma^2 N_n^j \log\log N_n^j}} = 1 \quad a.s.$$

and

$$\liminf \frac{S_n - n\mu}{\sqrt{2\sigma^2 N_n^j \log\log N_n^j}} = -1 \quad a.s.$$

since $N_n^j \to \infty$. Now we have

$$\frac{\log N_n^j}{\log n} \to \frac{\log n}{\log n} + \frac{\log \pi_j}{\log n} \to 1$$

and

$$\log\log N_n^j - \log\log n \to \log\left(1 + \frac{\log \pi_j}{\log n}\right) \to 0.$$

Using (5.15) and by the same argument as in the proof of Theorem 5.1 the desired result follows. □

To see that $\sigma^2 = \pi_j \mathrm{Var}(Y_1^j)$ is actually independent of π_j we need to consider the time-reversed Markov chain of the MRW defined as follows.

Definition 5.1. $\{\widehat{J}_n, n \geq 0\}$ is defined to be a time-reversed Markov chain induced by J if its transition probability measure is given by

$$\widehat{P}_{jk} = \frac{\pi_k}{\pi_j} P_{kj} \qquad (j, k \in E) \tag{5.18}$$

where P_{kj} is the transition matrix of J, $\{\pi_i\}$ is necessarily the stationary measure of both J and $\widehat{J}$.

Lemma 5.2. *We may express (5.12) in terms of the stationary distribution of J as follows.*

$$E_j(S_1^j \tau_1^j) = \frac{1}{\pi_j} \sum_{k \in E} \pi_k E_k(X_1)[1 + E_k(\widehat{\tau}_1^j)]. \tag{5.19}$$

Proof. In terms of the time-reversed Markov chain $\widehat{J}$, we have in view of Lemma 4.3 with the notation of Sec. 4.4

$$\sum_{m=0}^{\infty} mP_j\{\tau_1^j > m; J_m = k\} = \sum_{m=1}^{\infty} m\frac{\pi_k}{\pi_j}P_k\{\widehat{\tau}_1^j > m-1; \widehat{J}_m = j\}$$
$$= \frac{\pi_k}{\pi_j}E_k(\widehat{\tau}_1^j).$$

The result follows by substituting this into (5.11) and using the fact that $E_j(\tau_1^j) = 1/\pi_j$. □

Remark 5.2. Substituting (5.13),(5.14), and (5.19) into (5.7), we get

$$\sigma^2 = \sum_{k\in E}\pi_k E_k(X_1^2) + 2\sum_{i,k\neq j}\pi_i E_i(X_1; J_1 = k)$$
$$\cdot\left[\sum_{l\in E}E_l(X_1)\pi_l\left[E_k(\tau_1^j) + E_j(\tau_1^l) - E_k(\tau_1^l)\right]\right]$$
$$+\mu^2\left[1 + 2\sum_{k\neq j}\pi_k E_k(\tau_1^j)\right] - 2\mu\sum_{k\in E}\pi_k E_k(X_1)\left[1 + E_k(\widehat{\tau}_1^j)\right]. \quad (5.20)$$

It can be seen that σ^2 is actually independent of π_j and thus independent of j.

5.4 Application

In Prabhu and Tang [94] the Central Limit Theorem is applied to obtain the heavy traffic limit distribution of the waiting time of some Markov-modulated single server queueing systems. Here, we discuss another possible application in finance.

Suppose that the profits or losses from a single trading can be modeled by the increments of the MRW. This is plausible under discrete trading and when the prices of stocks, for example, depend on a number of different market conditions which in turn can be modeled by a Markov chain J. After identifying different types of market conditions with the elements of E and with sufficient data of the profits and losses trading under these conditions, one may use the method given in Karr [51] to estimate the transition measure Q of the MRW.

Using the estimated Q, we can use Theorem 5.1 to compute the probability of the long term gain or losses from the investment. From Theorem 5.2, we can deduce the long term bounds for the gain and losses. The upper bound which is equal to $\sqrt{2\sigma^2 n \log\log n}$ assuming that $\mu = 0$, may be used to devise some stopping rule. Moreover, as a future research direction, the above results will be useful for assessing the long term risk under Markovian trading environment and proposing the appropriate capital levels associated with various ruin probabilities.

Chapter 6

Markov-Additive Processes of Arrivals

In this chapter we address Markov-additive processes (MAPs) of arrivals, a class of Markov processes with important applications as arrival processes to queueing networks. An MAP of arrivals is as an MAP whose additive component takes values in the nonnegative (may be multidimensional) integers, so that its increments may be interpreted as corresponding to arrivals, the standard example being that of different classes of arrivals into a queueing system. Many popular arrivals processes are special cases of univariate MAPs of arrivals, a number of which are reviewed in the chapter.

For MAPs of arrivals, we investigate the lack of memory property, interarrival times, moments of the number of counts, and limit theorems for the number of counts. We then consider transformations of these processes that preserve the Markov-additive property, such as linear transformations, patching of independent processes, and linear combinations. Random time transformations are also investigated. Finally we consider secondary recordings that generate new arrival processes from an original MAP of arrivals; these include, in particular, marking, colouring and thinning. For Markov-Bernoulli recording the secondary process in each case turns out to be an MAP of arrivals.

In the next section we present a brief introduction to MAPs of arrivals, but before that we introduce some notation used in the chapter. We endow the vector space $\mathbf{R}^r$ with the usual componentwise addition of vectors and multiplication of a vector by a scalar, and denote the points $\mathbf{x} = (x_1, x_2, \ldots, x_r)$ of $\mathbf{R}^r$, $r \geq 2$, in boldface. In $\mathbf{R}^r$ we consider partial

---

[6]This chapter is an updated version of material included in the article by A. Pacheco and N.U. Prabhu, 'Markov-additive processes of arrivals', in J. H. Dshalalow (ed.), *Advances in Queueing: Theory, Methods, and Open Problems*, Probability and Stochastics Series (CRC, Boca Raton, FL), pp. 167–194, 1995, and contents are reproduced with permission of the publisher, CRC Press.

order relations $\leq$, $<$, $\geq$ and $>$ defined by

$$\mathbf{x} \leq \mathbf{y} \Leftrightarrow x_i \leq y_i,\ i = 1, 2, \ldots, r,$$

$\mathbf{x} < \mathbf{y} \Leftrightarrow \mathbf{x} \leq \mathbf{y}$ and $\mathbf{x} \neq \mathbf{y}$, $\mathbf{x} \geq \mathbf{y} \Leftrightarrow \mathbf{y} \leq \mathbf{x}$, and $\mathbf{x} > \mathbf{y} \Leftrightarrow \mathbf{y} < \mathbf{x}$. We denote by $\mathbf{e}$ a vector with unit elements and by $\mathbf{0}$ a vector of zeros, with dimensions being clear from the context. Given matrices $A = (a_{jk})$ and $B = (b_{jk})$ of the same order, we denote the *Schur* or *entry-wise multiplication* of A and B by $A \bullet B = (a_{jk}\, b_{jk})$. Moreover, given matrices $A = (a_{jk})_{F_1 \times F_2}$ and $B = (b_{il})_{F_3 \times F_4}$ we let

$$A \oplus B = (c_{(j,i)(k,l)})_{(F_1 \times F_3) \times (F_2 \times F_4)} = (a_{jk}\, \delta_{il} + \delta_{jk}\, b_{il})$$

denote the *Kronecker sum* of A and B.

6.1 Introduction

Since we will be interested in MAPs of arrivals with a multivariate additive component, we first recall the definition of MAP properly adapted to this multivariate setting. For that, we denote $\mathbf{R} = (-\infty, \infty)$, r a positive integer, and E a countable set.

Definition 6.1. A process $(\mathbf{X}, J) = \{(\mathbf{X}(t), J(t)),\ t \geq 0\}$ on the state space $\mathbf{R}^r \times E$ is an MAP if it satisfies the following two conditions:

(a) $(\mathbf{X}, J)$ is a Markov process;
(b) For $s, t \geq 0$, the conditional distribution of $(\mathbf{X}(s+t) - \mathbf{X}(s), J(s+t))$ given $(\mathbf{X}(s), J(s))$ depends only on $J(s)$.

In this framework, an MAP of arrivals is simply an MAP with the additive component taking values on the nonnegative (may be multidimensional) integers, i.e., $\mathbf{N}_+^r$ with $\mathbf{N}_+ = \{0, 1, 2, \ldots\}$. For clarity, in the chapter we consider time-homogeneous MAPs, for which the conditional distribution of $(\mathbf{X}(s+t) - \mathbf{X}(s), J(s+t))$ given $J(s)$ depends only on t. Our analysis of MAPs of arrivals is based on what we can say from the Markovian nature of these processes, rather than by looking at them as point processes.

As previously said, an MAP of arrivals $(\mathbf{X}, J)$ is simply an MAP taking values on $\mathbf{N}_+^r \times E$ and, as a consequence, the increments of $\mathbf{X}$ may then be associated to arrival events, with the standard example being that of different classes of arrivals into a queueing system. Accordingly, without loss of generality, we use the term *arrivals* to denote the events studied and make the interpretation

$$X_i(t) = \text{ total number of class } i \text{ arrivals in } (0, t]$$

for $i = 1, 2, \ldots, r$. We then call $\mathbf{X}$, the additive component in the terminology for MAPs, the *arrival component* of $(\mathbf{X}, J)$. We recall that it follows from the definition of MAP that J is a Markov chain and, following the terminology for MAPs, we call it the *Markov component* of $(\mathbf{X}, J)$.

Since an MAP of arrivals $(\mathbf{X}, J)$ is a Markov subordinator on $\mathbf{N}_+^r \times E$ it is a Markov chain. Moreover, since $(\mathbf{X}, J)$ is time-homogeneous it is characterized by its transition rates, which are translation invariant in the arrival component $\mathbf{X}$. Thus, it suffices to give, for $j, k \in E$ and $\mathbf{m}, \mathbf{n} \in \mathbf{N}_+^r$, the transition rate from $(\mathbf{m}, j)$ to $(\mathbf{m}+\mathbf{n}, k)$, which we denote simply (since the rate does not depend on $\mathbf{m}$) by $\lambda_{jk}(\mathbf{n})$.

Whenever the Markov component J is in state j, the following three types of transitions in $(\mathbf{X}, J)$ may occur with respective rates:

(a) Arrivals without change of state in J occur at rate $\lambda_{jj}(\mathbf{n})$, $\mathbf{n} > \mathbf{0}$;
(b) Change of state in J without arrivals occur at rate $\lambda_{jk}(\mathbf{0})$, $k \in E$, $k \neq j$;
(c) Arrivals with change of state in J occur at rate $\lambda_{jk}(\mathbf{n})$, $k \in E$, $k \neq j$, $\mathbf{n} > \mathbf{0}$.

We denote $\Lambda_{\mathbf{n}} = (\lambda_{jk}(\mathbf{n}))$, $\mathbf{n} \in \mathbf{N}_+^r$, and say that $(\mathbf{X}, J)$ is a *simple* MAP of arrivals if $\Lambda_{(n_1,\ldots,n_r)} = 0$ if $n_l > 1$ for some l. In case $\mathbf{X} \,(= X)$ has state space $\mathbf{N}_+$ we say that (X, J) is a *univariate* MAP of arrivals, an example of which is the Markov-Poisson process considered in Sec. 2.10, which is known as MMPP (Markov-modulated Poisson process).

The identification of the meaning of each of the above transitions is very important for applications. In particular, arrivals are clearly identified by transitions of types (a) and (c), so that we may talk about arrival epochs, interarrival times, and define complex operations like thinning of MAPs of arrivals. Moreover, the parametrization of these processes is simple and, by being Markov chains, they have "nice" structural properties. These are specially important computationally since it is a simple task to simulate Markov chains.

6.2 Univariate MAPs of Arrivals

In this section we present a brief account of the most common MAPs of arrivals investigated in the literature, namely univariate MAPs of arrivals (X, J) with finite Markov component, with emphasis on the evolution of the models proposed in the literature. We note that although there is a vast literature on univariate MAPs of arrivals, most of the time no connection to the theory on MAPs is made when addressing these processes.

The simplest univariate MAPs of arrivals correspond to the case where the Markov component has only one state. Then X is a continuous-time Markov process with stationary and independent increments on $\mathbf{N}_+$, and hence it is a compound Poisson process. In the terminology for MAPs of arrivals, the process is characterized by the rates $\Lambda_n = \lambda_n$ of arrivals of batches of size n ($n \geq 1$), and its generating function $G(z;t) = E[z^{X(t)}]$ is given by

$$G(z;t) = \exp\left\{-t \sum_{n\geq 1} \lambda_n (1 - z^n)\right\}. \tag{6.1}$$

The compound Poisson process allows only transitions of type (a) (arrivals without change of state in the Markov component) which are the only ones admissible anyway. If we leave this simple one-state case, and get into genuine examples of MAPs of arrivals, transitions of types (b) or (c) have to be allowed. Note that allowing only transitions of types (a) and (b) amounts to stating that only one of the components, X or J, changes state at a time; this is consistent with viewing X having a causal relationship with J.

To our knowdlege, the first defined (simple) MAP of arrivals, with transitions of types (a) and (b) was the MMPP, which is a Cox process with intensity rate modulated by a finite Markov chain. This process was given a definition as an MAP of arrivals by Prabhu [89] along the same lines of Sec. 2.10. The MMPP was first used in queueing models by P. Naor and U. Yechialli, followed by M.F. Neuts; a compilation of results and relevant references on this process is given by Fischer and Meier-Hellstern [37]. Part of the popularity of the MMPP is due to the fact that, as all Cox processes, it is more *bursty* than the Poisson process in the sense that the variance of the number of counts is greater than its mean (see, e.g., [56]), whereas for the Poisson process these quantities are equal.

A generalization of the MMPP where batches of arrivals are allowed, the so-called batch MMPP, has also been considered. This process may be constructed from m independent compound Poisson processes $Y_1, Y_2, \ldots, Y_m$ by observing the process Y_j whenever the Markov component J is in state j (this follows from Neveu's [74] characterization of Markov subordinators with finite Markov component with m states). This setting suggests immediately that the process has important properties, such as lack of memory, being closed under Bernoulli thinning and under superpositioning of independent processes, similar to those of the compound Poisson process and of the simple MMPP. The batch MMPP is an MAP of arrivals with the rate matrices Λ_n $(n \geq 1)$ being diagonal matrices. Its generating function is a natural matrix extension of the generating function of the compound Poisson process, namely

$$G(z;t) = \exp\left\{t\left[Q - \sum_{n\geq 1} \Lambda_n(1-z^n)\right]\right\} \tag{6.2}$$

where Q is the generator matrix of the Markov component J. In fact, all univariate MAPs of arrivals have this type of generating function, with the Λ_n not necessarily diagonal.

Transitions of types (a)-(c) were allowed for the first time by Rudemo [101], who considered a simple process in which, in addition to arrivals as in the MMPP, there may be arrivals at transition epochs of the Markov chain J, i.e. type (c) transitions, which are sometimes called Markov-modulated transitions in the literature. Explicitly, in Rudemo's model at the time a transition from state j to state k occurs in J, an arrival occurs if $(j,k) \in A$ and no arrival occurs otherwise, where $A \subseteq \{(i,l) \in E^2 : i \neq l\}$. This process is thus a simple MAP of arrivals for which Λ_1 is not necessarily diagonal. The restrictive feature of the process is that

$$\lambda_{jk}(0)\,\lambda_{jk}(1) = 0 \qquad (j,k \in E), \tag{6.3}$$

which may be inconvenient in modeling.

The restrictive condition (6.3) is not present in the *phase-type* (PH) renewal process introduced by M.F. Neuts (see [72]) as a generalization of the Poisson process, containing modifications of the Poisson process such as the Erlangian and hyperexponential arrival processes. Here the interarrival times have a phase-type distribution, i.e. they are identified as times until absorption in a finite state Markov chain with one absorbing state. The PH process may be used to model sources that are less bursty than Poisson sources.

The first defined univariate MAP of arrivals that achieves the full generality possible when the Markov component has finite state space was the versatile Markovian arrival process of Neuts, the so-called N-process. The definition of this process is a constructive one [72], and in the original formulation special care was taken to include arrivals of the same type as in both the MMPP and the PH renewal process. In addition to transitions of types (a)-(c) (in the set of transient states of a finite Markov chain), the process allows for phase-type arrivals with one absorbing state. As follows easily from Proposition 6.1 in Sec. 6.3, the inclusion of phase-type arrivals does not add any additional generality to the class of N-processes, and in fact makes the model overparametrized.

Lucantoni, Meier-Hellstern and Neuts [64] defined a second (simple) Markovian arrival process, which was generalized to allow for batch arrivals by Lucantoni [62], who named it batch Markovian arrival process (BMAP). Although the classes of BMAPs and N-processes are the same, the BMAP has a simpler definition, given in Remark 6.1 below, than the N-process. For a history of the BMAP and its applications, with special emphasis on matrix-geometric methods, and a very extensive list of references see [63]. The simple BMAP has been referred to as MAP; this is unfortunate since this process is just a particular case of a Markov additive process and MAP is also the standard acronym for Markov-additive processes (first used by Çinlar [28]).

Extending Neuts's idea of phase-type arrivals with one absorbing state, Machihara [66] considered a case of simple phase-type arrivals with more than one absorbing state (see Example 6.2 in this respect); this process and its extension to the case of batch arrivals proposed by Yamada and Machihara [118] are closely related to the compound phase-type Markov renewal process we discuss in Example 6.3. It is relevant the fact the *point processes of arrivals* (sequence of arrival epochs along with the corresponding batch sizes) associated with the classes of univariate MAPs of arrivals with finite Markov component, BMAPs, and Yamada and Machihara's arrival processes are the same, as shown in Proposition 6.2 below.

MAPs of arrivals arise in important applications, as components of more complex systems, specially in queueing and data communication models. They have been used to model overflow from trunk groups, superpositioning of packetized voice streams, and input to ATM networks, which are high-speed communication networks (see, e.g., [37, 63, 66, 72, 73, 111, 118] and references therein). They have also been used to establish queueing theoretic results and to investigate constructions on arrival streams (see e.g. [43, 63, 73] and references therein).

Queues with Markov-additive input have also been the subject of much study; see, e.g., [37, 63, 64, 66, 72, 93, 96, 111, 118] and references therein. The output from queues with Markov-additive input has also been considered; in general this process is not an MAP of arrivals. In particular, the output from an $MMPP/M/1$ queue is not an MAP of arrivals unless the input is Poisson [80]. Moreover, [111] investigates the extent to which the output process *conforms* to the input process generated by an MAP of arrivals for a high-speed communications network with the *leaky bucket* for *regulatory access mechanism* (or traffic shaping mechanism). The class of stationary univariate MAPs of arrivals with finite Markov component was shown to be dense in the family of all stationary point processes on $\mathbf{R}_+$ [13]; here an MAP is said to be stationary if its Markov component is stationary. This fact and the considerable tractability of MAPs of arrivals reinforce their importance in applications. For more details on applications of MAPs of arrivals, we refer the reader to the papers cited, and in particular to [63, 73, 96].

We have reviewed some univariate MAPs of arrivals (MAPs with the additive component taking values on $\mathbf{N}_+$) on continuous time. Although continuous-time models are the most common case in the literature, discrete time MAPs of arrivals have also received a great deal of attention. In particular, they have been applied to model packet based arrivals into transmission systems [15, 16, 77, 78, 102, 103].

6.3 MAPs of Arrivals

In this section we introduce additional notation for MAPs of arrivals and provide some examples and results. We let $(\mathbf{X}, J)$ denote a (time-homogeneous) MAP of arrivals on $\mathbf{N}_+^r \times E$. Due to its Markov-additive structure, state transitions in $(\mathbf{X}, J)$ may be identified by the states of J immediately before (j) and after (k) the transition occurs along with the corresponding observed increment in $\mathbf{X}$ $(\mathbf{n})$. Thus we characterize each transition in $(\mathbf{X}, J)$ by one element of the set

$$S_r(E) = \{(j, k, \mathbf{n}) \in E^2 \times \mathbf{N}_+^r : (\mathbf{n}, k) \neq (\mathbf{0}, j)\} \tag{6.4}$$

and each element $(j, k, \mathbf{n})$ of $S_r(E)$ has associated its respective nonnegative transition rate $\lambda_{jk}(\mathbf{n})$.

As regards the transition probability measure of $(\mathbf{X}, J)$, we have

$$\begin{aligned} P_{jk}(\mathbf{n}; h) &= P(\mathbf{X}(h) = \mathbf{n}, J(h) = k \mid J(0) = j) \\ &= \begin{cases} \lambda_{jk}(\mathbf{n})\, h + o(h) & (\mathbf{n}, k) \neq (\mathbf{0}, j) \\ 1 - \gamma_j\, h + o(h) & (\mathbf{n}, k) = (\mathbf{0}, j) \end{cases} \end{aligned} \tag{6.5}$$

with

$$\gamma_j = \sum_{\{(k,\mathbf{n}):(j,k,\mathbf{n}) \in S_r(E)\}} \lambda_{jk}(\mathbf{n}) < \infty \tag{6.6}$$

denoting the transition rate out of states for which the Markov component is in state j. The conditions (6.6) state that $(\mathbf{X}, J)$ is *stable*, and our setting implies that the infinitesimal generator of $(\mathbf{X}, J)$ is given by

$$\mathcal{A}f(\mathbf{m}, j) = \sum_{\{(k,\mathbf{n}):(j,k,\mathbf{n})\in S_r(E)\}} \lambda_{jk}(\mathbf{n}) \left[f(\mathbf{m}+\mathbf{n}, k) - f(\mathbf{m}, j)\right] \tag{6.7}$$

with f being a bounded real function on $\mathbf{N}_+^r \times E$. This shows again that the process is determined by the matrices $\Lambda_\mathbf{n} = (\lambda_{jk}(\mathbf{n}))$, $\mathbf{n} \in \mathbf{N}_+^r$, where we let $\lambda_{jk}(\mathbf{n}) = 0$ for $(j, k, \mathbf{n}) \notin S_r(E)$.

For convenience, we let

$$\Gamma = (\gamma_j\, \delta_{jk}), \quad \Lambda = (\lambda_{jk}) = \sum_{\mathbf{n}>\mathbf{0}} \Lambda_\mathbf{n}, \quad \Sigma = (\sigma_{jk}) = \sum_{\mathbf{n}\in\mathbf{N}_+^r} \Lambda_\mathbf{n}, \tag{6.8}$$

$$Q = (q_{jk}) = \Sigma - \Gamma, \quad Q^\mathbf{0} = (q_{jk}^\mathbf{0}) = \Lambda_\mathbf{0} - \Gamma. \tag{6.9}$$

By inspection of their entries, it is easy to see that:

- Λ is the matrix of transition rates in J associated with arrivals;
- Σ is the matrix of transition rates in J associated with both arrivals or non-arrivals; and
- Q is the generator matrix of the Markov component J.

As regards the matrix $Q^\mathbf{0}$ it has non-positive diagonal entries, nonnegative off-diagonal entries, and non-positive row sums. This implies in particular that $Q^\mathbf{0}$ is the transient part of the generator matrix of a Markov chain with one absorbing state. An inspection of its entries reveals that $Q^\mathbf{0}$ is the generator matrix associated with non-arrival transitions in $(\mathbf{X}, J)$.

As we can obtain the matrices $\{\Lambda_\mathbf{n}\}_{\mathbf{n}\geq\mathbf{0}}$ from the matrices $(Q, \{\Lambda_\mathbf{n}\}_{\mathbf{n}>\mathbf{0}})$, and vice-versa, we conclude that the process $(\mathbf{X}, J)$ is characterized by the set of matrices $(Q, \{\Lambda_\mathbf{n}\}_{\mathbf{n}>\mathbf{0}})$, and we say that $(\mathbf{X}, J)$ is an MAP of arrivals with $(Q, \{\Lambda_\mathbf{n}\}_{\mathbf{n}>\mathbf{0}})$-*source*. This term is a natural extension of the (Q, Λ)-source term commonly used in the literature on MMPPs, where $\Lambda = \Lambda_1$.

In applications it makes sense to assume that J is non-explosive, and, for a stable MAP of arrivals $(\mathbf{X}, J)$, this implies that $(\mathbf{X}, J)$ is also non-explosive, so that the number of arrivals in finite time intervals is a.s. finite, as shown in Lemma 6.1 below. Thus, in the rest of the paper we consider non-explosive MAPs of arrivals, which in particular are strong Markov. For simplicity, we impose a slightly stronger condition for $(\mathbf{X}, J)$, namely

$$\gamma = \sup_{j\in E} \gamma_j = \sup_{j\in E} \sum_{\{(k,\mathbf{n}):(j,k,\mathbf{n})\in S_r(E)\}} \lambda_{jk}(\mathbf{n}) < \infty. \tag{6.10}$$

However we note that (6.10) is not needed for some of the results we give, and that (6.10) holds trivially when E is finite.

Lemma 6.1. *If* $(\mathbf{X}, J)$ *is a stable MAP of arrivals then* $(\mathbf{X}, J)$ *is explosive if and only if* J *is explosive.*

Proof. Since J makes no more changes of state than $(\mathbf{X}, J)$, it follows trivially that if J is explosive then $(\mathbf{X}, J)$ is also explosive. Conversely, if J is non-explosive then in a finite time interval J changes state only a finite number of times a.s., and during a subinterval in which J is in the same state $\mathbf{X}$ changes state a finite number of times a.s., since $(\mathbf{X}, J)$ is stable. This leads to the conclusion that in finite time $(\mathbf{X}, J)$ changes state a finite number of times a.s., i.e. if J is non-explosive then $(\mathbf{X}, J)$ is also non-explosive, which completes the proof. □

For $s, t \geq 0$, $j, k \in E$, $\mathbf{n} \in \mathbf{N}_+^r$ the transition probability measure of $(\mathbf{X}, J)$ is such that

$$P_{jk}(\mathbf{n}; t) \geq 0, \quad P_{jk}(\mathbf{n}; 0) = \delta_{(\mathbf{0},j)(\mathbf{n},k)}, \quad \sum_{k \in E} P_{jk}(\mathbf{N}_+^r; t) = 1 \tag{6.11}$$

and the Chapman-Kolmogorov equations are

$$P_{jk}(\mathbf{n}; t+s) = \sum_{l \in E} \sum_{\mathbf{0} \leq \mathbf{m} \leq \mathbf{n}} P_{jl}(\mathbf{m}; t)\, P_{lk}(\mathbf{n} - \mathbf{m}; s). \tag{6.12}$$

As regards the transition probability measure of the Markov component J,

$$\pi_{jk}(t) = P\{J(t) = k \mid J(0) = j\}, \tag{6.13}$$

we have $\Pi(t) = (\pi_{jk}(t)) = (P_{jk}(\mathbf{N}_+^r; t))$, so that, in view of (6.5),

$$\pi_{jk}(h) = \sigma_{jk} h + o(h) \quad (h > 0,\, j \neq k).$$

This shows that Q is the generator matrix of the Markov component J of an MAP of arrivals with $(Q, \{\Lambda_{\mathbf{n}}\}_{\mathbf{n} > \mathbf{0}})$-source.

In the rest of the section we give some examples of MAPs of arrivals.

Example 6.1. (Arrivals, departures and overflow from a network).
Suppose we have a Markovian network with r nodes and let $J_i(t)$, $1 \leq i \leq r$, be the number of units at node i at time t, and $J = (J_1, J_2, \ldots, J_r)$. In addition, suppose that nodes $j_1, j_2, \ldots, j_s$ $(s \leq r)$ have finite capacity while the rest have infinite capacity. Let $Y_i(t)$ denote the number of external units which entered node i, $Z_i(t)$ denote the number of units which left the system from node i, and $W_l(t)$ denote the overflow at node l by time t, and let $\mathbf{X} = (\mathbf{Y}, \mathbf{Z}, \mathbf{W})$ with

$$\mathbf{Y} = (Y_1, Y_2, \ldots, Y_r), \quad \mathbf{Z} = (Z_1, Z_2, \ldots, Z_r), \quad \mathbf{W} = (W_1, W_2, \ldots, W_s).$$

The process $(\mathbf{X}, J)$ is an MAP of arrivals on a subset of $\mathbf{N}_+^{3r+s}$. This holds even with batch input and state dependent input, output or routing rates. We could also include arrival components counting the number of units going from one set of nodes to another.

The fact that $(\mathbf{X}, J)$ is an MAP holds in particular for the queueing networks with dependent nodes and concurrent movements studied by Serfozo [104]. Networks are systems with inherent dependencies, which may be either outside the control of the manager of the system or introduced by the manager. Dependencies may be introduced to avoid congestion, balance the workload at nodes, increase the throughput, etc.; thus the study of queueing networks with dependencies is very important for applications.

Example 6.2. (Compound phase-type (CPH) arrival processes). Consider a continuous-time Markov chain $J^\star$ on $\{1, 2, \ldots, m, m+1, \ldots, m+r\}$ with stable and conservative infinitesimal generator matrix

$$Q^\star = \begin{bmatrix} Q^0 & \Lambda^\star \\ 0 & 0 \end{bmatrix} \tag{6.14}$$

with Q^0 being a $m \times m$ matrix, so that states $m+1, \ldots, m+r$ are absorbing. After absorption into state $m+l$ the Markov chain is instantaneously restarted (independently of previous restartings) into transient state k with probability α_{lk}; moreover, if absorption is from state j, then associate with it an arrival of a batch of size $n > 0$ (independently of the size of other batches) with probability $p_{jl}(n)$.

An instance of this model was used by Machihara [66] to model service interruptions in a queueing system, where interruptions are initiated at absorption epochs. The author viewed these interruption epochs as the arrival epochs of phase-type Markov renewal customers. If the service interruptions are due to physical failure with the absorption state indicating the type of failure, then it becomes important to record the type of failure (absorption state) which occurs at each failure time (absorption epoch). Accordingly, we let $X_l(t)$, $1 \leq l \leq r$ be the number of arrivals associated with absorptions into state $m+l$ in $(0, t]$, and let J be the Markov chain on states $E = \{1, 2, \ldots, m\}$ obtained by carrying out the described instantaneous restartings after absorptions and requiring the sample functions of the resulting process to be right-continuous. If we let

$$\alpha = (\alpha_{lk}), \quad P_n = (p_{jl}(n)), \quad \Psi_n = (\psi_{jl}(n)) = \Lambda^\star \bullet P_n \quad (n \geq 1)$$

then it is easy to see that J has stable and conservative infinitesimal generator matrix

$$Q = Q^0 + \sum_{n \geq 1} \Psi_n \alpha. \tag{6.15}$$

Moreover, if we let $\mathbf{X} = (X_1, X_2, \ldots, X_r)$, then $(\mathbf{X}, J)$ is an MAP of arrivals on $\mathbf{N}_+^r \times E$ with $(Q, \{\Lambda_\mathbf{n}\}_{\mathbf{n}>\mathbf{0}})$-source, where

$$\lambda_{jk}(\mathbf{n}) = \begin{cases} \psi_{jl}(n_l)\,\alpha_{lk} & \text{if } n_l > 0 \text{ and } n_p = 0,\, p \neq l \\ 0 & \text{otherwise.} \end{cases}$$

Note that arrivals occur only in one of the coordinates of the arrival component at each time. Similarly, if we let $X(t)$ denote the total number of arrivals in $(0, t]$, then the process (X, J) is an univariate MAP of arrivals with $(Q, \{\Psi_n \alpha\}_{n>0})$-source. We call (X, J) a CPH arrival process with representation $(Q^0, \Lambda^\star, \{P_n\}_{n>0}, \alpha)$.

For control of the system it is important to know which types of failures occur most frequently in order to minimize the loss due to service interruptions, for which we need to consider the process $(\mathbf{X}, J)$ instead of the CPH arrival process (X, J). This shows that multivariate arrival components may be needed for an appropriate study of some systems. The use of (X, J) is justified only if different failures produce similar effects and have approximately equal costs.

Remark 6.1. If (Y, J) is an MAP of arrivals on $\mathbf{N}_+ \times \{1, 2, \ldots, m\}$ with $(Q, \{\Lambda_n\}_{n>0})$-source, and we use the lexicographic ordering of the states of (Y, J), then (Y, J) has infinitesimal generator matrix with upper triangular block structure

$$\begin{bmatrix} Q^0 & \Lambda_1 & \Lambda_2 & \Lambda_3 & \cdots \\ 0 & Q^0 & \Lambda_1 & \Lambda_2 & \cdots \\ 0 & 0 & Q^0 & \Lambda_1 & \cdots \\ \cdots & \cdots & \cdots & \cdots & \cdots \end{bmatrix}. \tag{6.16}$$

A BMAP with representation $\{D_k,\, k \geq 0\}$ has generator matrix of the form (6.16) with $D_0 = Q^0$ and $D_n = \Lambda_n$, $n \geq 1$. For the BMAP it is sometimes assumed that Q^0 is invertible and Q is irreducible, but these conditions are not essential.

Proposition 6.1. *The class of CPH arrival processes is equivalent to the class of univariate MAPs of arrivals with finite Markov component (i.e. the class of BMAPs).*

Proof. A CPH arrival process (X, J) with representation $(Q^0, \Lambda^\star, \{P_n\}_{n>0}, \alpha)$ is a univariate MAP of arrivals with $(Q, \{(\Lambda^\star \bullet P_n)\alpha\}_{n>0})$-source, as shown in Example 6.2, and has finite Markov component.

Conversely, suppose we are given an MAP of arrivals (Y, J) on $\mathbf{N}_+ \times \{1, 2, \ldots, m\}$ with $(Q, \{\Lambda_n\}_{n>0})$-source. For $n > 0$ we can obtain matrices $P_n = (p_{jk}(n))$ such that $\Lambda_n = \Lambda \bullet P_n$ with $\{p_{jk}(n),\, n > 0\}$ being a probability function, where we recall that $\Lambda = \sum_{n>0} \Lambda_n$. The process (Y, J) is then indistinguishable from a CPH arrival process with representation $(Q^0, \Lambda, \{P_n\}_{n>0}, I)$. □

Although the use of phase-type arrivals may be important from an operational point of view, in the context of univariate MAPs of arrivals with finite Markov component everything that may be accomplished with phase-type arrivals may be achieved without phase-type arrivals, and vice-versa, as Proposition 6.1 shows. This suggests some care in the definition of univariate arrival processes which have (conceptually) both phase-type and non-phase-type arrivals. A very good example of the need for a careful definition of these processes is given by the N-process. Since the N-process allows for changes of state of types (a)-(c) (not associated with phase-type arrivals) in their full generality in case the Markov component is finite, the addition of phase-type arrivals makes the model overparametrized. If, in practice, the process studied suggests the consideration of both phase-type arrivals and arrivals of types (a)-(c), then the individualization of these two classes of arrivals may be done easily using MAPs of arrivals with multivariate arrival component.

6.4 Some Properties of MAPs of Arrivals

For MAPs of arrivals we investigate in this section the partial lack of memory property, interarrival times, moments of the number of counts, and a strong law of large numbers. For that it is useful to introduce the following notation. For a countable set F we let $B(F)$ be the space of bounded real functions on F, $l_\infty(F)$ be the Banach space of real sequences $a = (a_j)_{j \in F}$ with the norm $\|a\| = \sup_{j \in F} |a_j|$ and $B(l_\infty(F))$ be the space of bounded linear operators on $l_\infty(F)$, an element W of which may be identified by a matrix $W = (w_{jk})_{j,k \in F}$ with norm

$$\|W\| = \sup_{j \in F} \sum_{k \in F} |w_{jk}|.$$

Following a similar approach to that presented in Sec. 2.9 to derive Theorem 2.11, we have the following result.

Lemma 6.2. *If $A \in B(l_\infty(E))$, the matrix differential equation $R'(t) = R(t)\,A$ with the condition $R(0) = I$ has $R(t) = e^{tA}$ as its unique solution in $B(l_\infty(E))$.*

We consider first the partial lack of memory property. Since the arrival component $\mathbf{X}$ has non-decreasing sample functions (i.e. $\mathbf{X}(t) \geq \mathbf{X}(s)$ a.s. for $0 \leq s \leq t$), the epoch of first arrival becomes

$$T^\circ = \inf\{t : \mathbf{X}(t) > \mathbf{0}\}. \tag{6.17}$$

Theorem 6.1. *If $(\mathbf{X}, J)$ is an MAP of arrivals on $\mathbf{N}_+^r \times E$ with $(Q, \{\Lambda_\mathbf{n}\}_{\mathbf{n}>\mathbf{0}})$-source, then:*

(a) $T^\circ > t \Leftrightarrow \mathbf{X}(t) = \mathbf{0}$ *a.s.*
(b) T° *is a stopping time.*
(c) (**Partial lack of memory property**).
For $t \in \mathbf{R}_+$ and $j, k \in E$, denote

$$U^\mathbf{0}_{jk}(t) = P\{T^\circ > t,\, J(t) = k \mid J(0) = j\} \tag{6.18}$$

and $U^\mathbf{0}(t) = (U^\mathbf{0}_{jk}(t))$. Then, the family of matrices $\{U^\mathbf{0}(t),\, t \in \mathbf{R}_+\}$ forms a semigroup, i.e.

$$U^\mathbf{0}(t+s) = U^\mathbf{0}(t)\,U^\mathbf{0}(s) \quad (t, s \in \mathbf{R}_+). \tag{6.19}$$

(d) *Moreover, $U^\mathbf{0}(t) = e^{tQ^\mathbf{0}}$, so that (6.19) reads*

$$e^{(t+s)Q^\mathbf{0}} = e^{tQ^\mathbf{0}} e^{sQ^\mathbf{0}} \quad (t, s \in \mathbf{R}_+). \tag{6.20}$$

Proof. The statement (a) follows easily from the fact that $(\mathbf{X}, J)$ is a Markov subordinator, and the statement (b) is an immediate consequence of (a).
(c). From (a), $U^\mathbf{0}_{jk}(u) = P_{jk}(\mathbf{0}; u)$ for $j, k \in E$ and $u \in \mathbf{R}_+$. From the Chapman-Kolmogorov equations (6.12) we find that the semigroup property (6.19) holds.
(d). From (6.18) and (6.5), using (a), we have for $j, k \in E$

$$U^\mathbf{0}_{jk}(h) = \begin{cases} \lambda_{jk}(\mathbf{0})\,h + o(h) & j \neq k \\ 1 - \gamma_j\,h + o(h) & j = k \end{cases} \tag{6.21}$$

so that $\{U^\mathbf{0}(t),\, t \geq 0\}$ is a continuous semigroup with infinitesimal generator matrix $Q^\mathbf{0}$. Moreover $U^\mathbf{0}(0) = I$, $[U^\mathbf{0}(t)]' = U^\mathbf{0}(t)\,Q^\mathbf{0}$, and $\|Q^\mathbf{0}\| = \|\Lambda_\mathbf{0} - \Gamma\| \leq \|\Lambda_\mathbf{0}\| + \|\Gamma\| \leq 2\gamma < \infty$, which implies that $U^\mathbf{0}(t) = e^{tQ^\mathbf{0}}$, in view of Lemma 6.2. □

The partial lack of memory property has been investigated for the MMPP by Prabhu [89] as a natural extension of the well known lack of memory property of the Poisson process. From (6.18), we see that $Q^{\mathbf{0}}$ is the generator matrix of the transitions in J not associated with arrivals in $(\mathbf{X}, J)$. We note that Q^0 was used as a building block of the BMAP, the CPH arrival process and Yamada and Machihara's arrival process, to be discussed in Example 6.3.

We define the successive arrival epochs of the MAP of arrivals $(\mathbf{X}, J)$

$$T_p^\circ = \inf\{t : \mathbf{X}(t) > \mathbf{X}(T_{p-1}^\circ)\} \quad (p \geq 1) \tag{6.22}$$

where $T_0^\circ = 0$, so that $T_1^\circ = T^\circ$. Owing to the presence of the Markov process J, we expect the interarrival times $T_p^\circ - T_{p-1}^\circ$ to be Markov-dependent.

Theorem 6.2 (Interarrival times). *The process $\{(T_p^\circ, \mathbf{X}_p^\circ, J_p^\circ),\, p \geq 0\}$, with $\mathbf{X}_p^\circ = \mathbf{X}(T_p^\circ)$ and $J_p^\circ = J(T_p^\circ)$, is an MRP whose one-step transition probability density is given by*

$$V(t, \mathbf{n}) = (v_{jk}(t, \mathbf{n})) = e^{tQ^{\mathbf{0}}} \Lambda_{\mathbf{n}}. \tag{6.23}$$

Proof. Since $(\mathbf{X}, J)$ is non-explosive it is a strong Markov process. Now since

$$\begin{cases} T_{p+1}^\circ - T_p^\circ = \inf\{t - T_p^\circ : \mathbf{X}(t) > \mathbf{X}(T_p^\circ)\} \\ \mathbf{X}_{p+1}^\circ - \mathbf{X}_p^\circ = \mathbf{X}(T_{p+1}^\circ) - \mathbf{X}(T_p^\circ) \end{cases}$$

we see that given $J_0^\circ, (T_1^\circ, \mathbf{X}_1^\circ, J_1^\circ), \ldots, (T_p^\circ, \mathbf{X}_p^\circ, J_p^\circ)$, the distribution of $(T_{p+1}^\circ - T_p^\circ, \mathbf{X}_{p+1}^\circ - \mathbf{X}_p^\circ, J_{p+1}^\circ)$ depends only on J_p° and is the same as that of $(T_1^\circ, \mathbf{X}_1^\circ, J_1^\circ)$ given J_0°. This implies that $\{(T_p^\circ, \mathbf{X}_p^\circ, J_p^\circ),\, p \geq 0\}$ is an MRP. Moreover, since the probability of two or more transitions in $(\mathbf{X}, J)$ in a time interval of length h is of order $o(h)$ we have for the one-step transition probability density of the MRP

$$\begin{aligned} \big(P\{T_{p+1}^\circ - T_p^\circ \in (t, t+dt],\, \mathbf{X}_{p+1}^\circ - \mathbf{X}_p^\circ = \mathbf{n},\, J_{p+1}^\circ = k \mid J_p^\circ = j\}\big) \\ = \left(\sum_{l \in E} P_{jl}(\mathbf{0}; t) P_{lk}(\mathbf{n}; dt)\right) + o(dt) \\ = U^{\mathbf{0}}(t) \Lambda_{\mathbf{n}} dt + o(dt), \end{aligned}$$

which leads to (6.23), in view of Theorem 6.1 (d). □

Interarrival times have received much consideration in the applied literature. Some authors have used the MRP $\{(T_p^\circ, \mathbf{X}_p^\circ, J_p^\circ)\}$ as defined in Theorem 6.2 to characterize the arrival process of special cases of MAPs of arrivals (e.g. [73]). From Theorem 6.2, we can define a semi-Markov process $(\mathbf{X}^\star, J^\star)$ by letting $(\mathbf{X}^\star, J^\star)(t) = (\mathbf{X}_p^\circ, J_p^\circ)$, for $T_p^\circ \leq t < T_{p+1}^\circ$, but this process gives only partial information on the Markov component of the MAP of arrivals $(\mathbf{X}, J)$.

Example 6.3. (Compound phase-type Markov renewal process). Consider a CPH arrival process (X, J) with representation $(Q^0, \Lambda^\star, \{P_n\}_{n>0}, \alpha)$, as explained in Example 6.2. We let $\{T_p^\circ\}$ be the epochs of increments in X and denote $J_p^\circ = J(T_p^\circ)$ and $X_p^\circ = X(T_p^\circ)$. Then, by Theorem 6.2, the process $\{(T_p^\circ, X_p^\circ, J_p^\circ),\ p \in \mathbf{N}_+\}$ is an MRP with transition probability density

$$V(t,n) = (v_{jk}(t,n)) = e^{tQ^0}(\Lambda^\star \bullet P_n)\,\alpha.$$

We shall call this process compound phase-type Markov renewal process (CPH-MRP) with representation $(Q^0, \Lambda^\star, \{P_n\}_{n>0}, \alpha)$. Its associated point process of arrivals is $\{(T_p^\circ, X_p^\circ - X_{p-1}^\circ),\ p \geq 1\}$; this process has information about the arrival epochs and the batch sizes of arrivals, but not of the Markov component of the MRP.

Closely related to this is the process defined for simple arrivals by Machihara [66] and for the general case by Yamada and Machihara [118]; in fact, in the context of Example 6.2, the components T_p° and X_p° of the two processes are the same, while the Markov component is $J^\star$, where $J_p^\star$ represents the state the p-th absorption occurs into. This MRP $\{(T_p^\circ, X_p^\circ, J_p^\star)\}$ has transition probability density

$$V^\star(t,n) = \alpha e^{tQ^0}(\Lambda^\star \bullet P_n)$$

and the authors state that it has the representation $(\alpha, Q^0, \Lambda^\star, \{P_n\}_{n>0})$. Yamada an Machihara assume that Q^0 is invertible, but this condition is not essential. We note that unless the initial distribution π° of J° and $\pi^\star$ of $J^\star$ are chosen so that $\pi^\circ = \pi^\star\alpha$ the point processes of arrivals associated with $\{(T_p^\circ, X_p^\circ, J_p^\circ)\}$ and $\{(T_p^\circ, X_p^\circ, J_p^\star)\}$ need not be stochastically equivalent.

Our definition of CPH-MRPs does not coincide with the definitions in [66], [73], and [118]. In [73] PH-MRPs are MRPs with interarrival times with a phase-type distribution.

We now state and prove a result for univariate MAPs of arrivals that we have mentioned at the end of Sec. 6.2.

Proposition 6.2. *The classes of point processes of arrivals associated with CPH-MRPs, Yamada and Machihara's processes, univariate MAPs of arrivals with finite Markov component, and BMAPs are equal.*

Proof. From Proposition 6.1 it is clear that the classes of point processes of arrivals associated with CPH-MRPs, univariate MAPs of arrivals with finite Markov component, and BMAPs are equal. We now show that the class of point processes of arrivals associated with Yamada and Machihara's processes and univariate MAPs of arrivals with finite Markov component are the same. Suppose we are given an Yamada and Machihara's process with representation $(\alpha, Q^0, \Lambda^\star, \{P_n\}_{n>0})$ and initial distribution $\pi^\star$ for the Markov component. From Examples (6.2) and (6.3), it is easy to see that this process has the same associated point process of arrivals as an univariate MAP of arrivals (X, J) with $(Q^0 + \sum_{n\geq 1}(\Lambda^\star \bullet P_n)\alpha, \{(\Lambda^\star \bullet P_n)\alpha\}_{n>0})$-source and initial distribution $\pi^\star\alpha$ for J. Conversely, an univariate MAP of arrivals with finite Markov component, $(Q, \{\Lambda_n\}_{n>0})$-source and initial distribution π for J has the same associated point process of arrivals as an Yamada and Machihara's process with representation $(I, Q^0, \Lambda, \{P_n\}_{n>0})$ and initial distribution π for the Markov component, where P_n is chosen so that $\Lambda_n = \Lambda \bullet P_n$. □

By considering the process $(\mathbf{X}, J)$ over the time intervals $(0, t]$ and $(t, t+dt]$ and using (6.12) and (6.5), we obtain the relation

$$P_{jk}(\mathbf{n}; t+dt) = P_{jk}(\mathbf{n}; t)[1 - \gamma_k\, dt] + \sum_{\mathbf{0}\leq\mathbf{m}\leq\mathbf{n}} \sum_{l\in E} P_{jl}(\mathbf{m}; t)\, \lambda_{lk}(\mathbf{n}-\mathbf{m})\, dt + o(dt)$$

for $j, k \in E$ and $\mathbf{n} \in \mathbf{N}_+^r$. As a result

$$\frac{\partial}{\partial t} P_{jk}(\mathbf{n}; t) = -\gamma_k\, P_{jk}(\mathbf{n}; t) + \sum_{\mathbf{0}\leq\mathbf{m}\leq\mathbf{n}} \sum_{l\in E} P_{jl}(\mathbf{m}; t)\, \lambda_{lk}(\mathbf{n}-\mathbf{m})$$

which leads to the following matrix differential equation:

$$\frac{\partial}{\partial t} P(\mathbf{n}; t) = -P(\mathbf{n}; t)\,\Gamma + \sum_{\mathbf{0}\leq\mathbf{m}\leq\mathbf{n}} P(\mathbf{m}; t)\,\Lambda_{\mathbf{n}-\mathbf{m}}. \tag{6.24}$$

To solve this, we introduce the generating function matrix

$$G(\mathbf{z}; t) = (G_{jk}(\mathbf{z}; t)) = \left(\sum_{\mathbf{n}\in\mathbf{N}_+^r} P_{jk}(\mathbf{n}; t)\, \mathbf{z}^{\mathbf{n}} \right) \tag{6.25}$$

for $\mathbf{z} \in \mathbf{R}_+^r$ such that $\mathbf{0} \leq \mathbf{z} \leq \mathbf{e}$, with $\mathbf{z}^{\mathbf{n}} = \prod_{i=1}^r z_i^{n_i}$. Then, (6.24) leads to

$$\frac{\partial}{\partial t} G(\mathbf{z}; t) = G(\mathbf{z}; t)\,[\Phi(\mathbf{z}) - \Gamma] \tag{6.26}$$

with

$$\Phi(\mathbf{z}) = \sum_{\mathbf{n}\in\mathbf{N}_+^r} \Lambda_{\mathbf{n}}\, \mathbf{z}^{\mathbf{n}}. \tag{6.27}$$

Theorem 6.3. *The generating function matrix of an MAP of arrivals with* $(Q, \{\Lambda_{\mathbf{n}}\}_{\mathbf{n}>\mathbf{0}})$*-source is given by*

$$G(\mathbf{z};t) = \exp\{t[\Phi(\mathbf{z}) - \Gamma]\} \tag{6.28}$$

$$= \exp\left\{t\left[Q - \sum_{\mathbf{n}>\mathbf{0}} \Lambda_{\mathbf{n}}(1 - \mathbf{z}^{\mathbf{n}})\right]\right\} \tag{6.29}$$

and $\Pi(t) = e^{tQ}$.

Proof. Since $\Phi(\mathbf{z}) - \Gamma = Q - \sum_{\mathbf{n}>\mathbf{0}} \Lambda_{\mathbf{n}}(1 - \mathbf{z}^{\mathbf{n}})$, (6.28) and (6.29) are equivalent. Moreover, in view of (6.26) and Lemma 6.2, to prove (6.28) it suffices to show that $\|\Phi(\mathbf{z}) - \Gamma\| < \infty$. But since $\|\Gamma\| = \|\Sigma\| = \gamma$ we have

$$\|\Phi(\mathbf{z}) - \Gamma\| \le \|\Phi(\mathbf{z})\| + \|\Gamma\| = \|\sum_{\mathbf{n}\in\mathbf{N}_+^r} \Lambda_{\mathbf{n}}\, \mathbf{z}^{\mathbf{n}}\| + \gamma$$

$$\le \|\sum_{\mathbf{n}\in\mathbf{N}_+^r} \Lambda_{\mathbf{n}}\| + \gamma = \|\Sigma\| + \gamma = 2\gamma < \infty.$$

This proves (6.28). The fact that $\Pi(t) = e^{tQ}$ follows from (6.29) by letting $\mathbf{z} \uparrow \mathbf{e}$. □

We will next capitalize on the previous result to derive moments of the number of counts of MAPs of arrivals. These have been considered for some particular cases of MAPs of arrivals by different authors, namely [37, 71, 72, 76, 89]. For a linear combination with nonnegative integer coefficients $Y = \alpha\mathbf{X}$ of $\mathbf{X}$ and $p \in \mathbf{N}_+$, we let

$$I\!E_p^Y(t) = (E[Y^p(t); J(t) = k \mid J(0) = j]) \tag{6.30}$$

$$= \left(\sum_{\mathbf{n}\in\mathbf{N}_+^r} (\alpha\mathbf{n})^p P_{jk}(\mathbf{n};t)\right) \tag{6.31}$$

and

$$\overline{\Sigma}_p^Y = \sum_{\mathbf{n}\in\mathbf{N}_+^r} (\alpha\mathbf{n})^p \Lambda_{\mathbf{n}}. \tag{6.32}$$

Theorem 6.4. *For* $1 \le i \le r$,

$$I\!E_1^{X_i}(t) = \int_0^t \Pi(s)\, \overline{\Sigma}_1^{X_i}\, \Pi(t-s)\, ds \tag{6.33}$$

and

$$\mathbb{E}_2^{X_i}(t) = \int_0^t \left[2\mathbb{E}_1^{X_i}(s)\,\overline{\Sigma}_1^{X_i} + \Pi(s)\,\overline{\Sigma}_2^{X_i}\right]\,\Pi(t-s)\,ds\,. \tag{6.34}$$

Proof. From (6.26), we have for $\mathbf{z} \in \mathbf{R}_+^r$ such that $\mathbf{0} \le \mathbf{z} \le \mathbf{e}$,

$$\begin{aligned}\frac{\partial}{\partial t}\frac{\partial}{\partial z_i}\,G(\mathbf{z};t) &= \frac{\partial}{\partial z_i}\frac{\partial}{\partial t}\,G(\mathbf{z};t) = \frac{\partial}{\partial z_i}\,\{G(\mathbf{z};t)\,[\Phi(\mathbf{z}) - \Gamma]\} \\ &= \left[\frac{\partial}{\partial z_i}G(\mathbf{z};t)\right][\Phi(\mathbf{z}) - \Gamma] + G(\mathbf{z};t)\frac{\partial}{\partial z_i}\Phi(\mathbf{z}).\end{aligned} \tag{6.35}$$

Making $\mathbf{z} = \mathbf{e}$ in the previous equation and using the fact that

$$\left[\frac{\partial}{\partial z_i}G(\mathbf{z};t)\right]_{\mathbf{z}=\mathbf{e}} = \mathbb{E}_1^{X_i}(t), \qquad \left[\frac{\partial}{\partial z_i}\Phi(\mathbf{z})\right]_{\mathbf{z}=\mathbf{e}} = \overline{\Sigma}_1^{X_i} \tag{6.36}$$

along with

$$G(\mathbf{e};t) = \Pi(t) = e^{t\,Q}, \qquad \Phi(\mathbf{e}) - \Gamma = \Sigma - \Gamma = Q \tag{6.37}$$

we find that $\mathbb{E}_1^{X_i}(t)$ satisfies the differential equation

$$\frac{\partial}{\partial t}\mathbb{E}_1^{X_i}(t) - \mathbb{E}_1^{X_i}(t)\,Q = e^{t\,Q}\,\overline{\Sigma}_1^{X_i} \tag{6.38}$$

which reads as $f'(t) = e^{t\,Q}\overline{\Sigma}_1^{X_i}e^{-t\,Q}$, with $f(t) = \mathbb{E}_1^{X_i}(t)e^{-t\,Q}$. As $f(0) = 0$, then

$$f(t) = \mathbb{E}_1^{X_i}(t)\,e^{-t\,Q} = \int_0^t f'(s)\,ds = \int_0^t e^{s\,Q}\overline{\Sigma}_1^{X_i}e^{-s\,Q}\,ds$$

which leads to (6.33) by postmultiplying by $e^{t\,Q}$ and taking into account that $\Pi(u) = e^{u\,Q}$, $u \ge 0$.

Proceeding as we have done to obtain (6.35), we are lead to

$$\begin{aligned}\frac{\partial}{\partial t}\frac{\partial^2}{\partial z_i^2}\,G(\mathbf{z};t) &= \left[\frac{\partial^2}{\partial z_i^2}G(\mathbf{z};t)\right][\Phi(\mathbf{z}) - \Gamma] \\ &\quad + 2\,\frac{\partial}{\partial z_i}G(\mathbf{z};t)\frac{\partial}{\partial z_i}\Phi(\mathbf{z}) + G(\mathbf{z};t)\frac{\partial^2}{\partial z_i^2}\Phi(\mathbf{z}).\end{aligned} \tag{6.39}$$

Making $\mathbf{z} = \mathbf{e}$ in the previous equation and using the fact that

$$\left[\frac{\partial^2}{\partial z_i^2}G(\mathbf{z};t)\right]_{\mathbf{z}=\mathbf{e}} = \mathbb{E}_2^{X_i}(t) - \mathbb{E}_1^{X_i}(t), \quad \left[\frac{\partial^2}{\partial z_i^2}\Phi(\mathbf{z})\right]_{\mathbf{z}=\mathbf{e}} = \overline{\Sigma}_2^{X_i} - \overline{\Sigma}_1^{X_i} \tag{6.40}$$

along with (6.36) and (6.37), we find that the function

$$g(t) = \mathbb{E}_2^{X_i}(t) - \mathbb{E}_1^{X_i}(t) \quad (t \ge 0)$$

satisfies the differential equation

$$\frac{\partial}{\partial t}g(t) - g(t)\,Q = 2\,I\!E_1^{X_i}(t)\overline{\Sigma}_1^{X_i} + e^{t\,Q}\left[\overline{\Sigma}_2^{X_i} - \overline{\Sigma}_1^{X_i}\right] \tag{6.41}$$

which is equivalent to

$$\frac{\partial}{\partial t}I\!E_2^{X_i}(t) - I\!E_2^{X_i}(t)\,Q = 2\,I\!E_1^{X_i}(t)\overline{\Sigma}_1^{X_i} + e^{t\,Q}\,\overline{\Sigma}_2^{X_i} \tag{6.42}$$

in view of (6.38) and the fact that

$$\frac{\partial}{\partial t}g(t) = \frac{\partial}{\partial t}I\!E_2^{X_i}(t) - \frac{\partial}{\partial t}I\!E_1^{X_i}(t).$$

Making $h(t) = I\!E_2^{X_i}(t)\,e^{-t\,Q}$, (6.42) reads as

$$h'(t) = \left[2\,I\!E_1^{X_i}(t)\overline{\Sigma}_1^{X_i} + e^{t\,Q}\,\overline{\Sigma}_2^{X_i}\right]e^{-t\,Q}$$

so that, as $h(0) = 0$,

$$h(t) = \int_0^t h'(s)\,ds = \int_0^t \left\{2\,I\!E_1^{X_i}(s)\overline{\Sigma}_1^{X_i} + e^{s\,Q}\,\overline{\Sigma}_2^{X_i}\right\}e^{-s\,Q}\,ds$$

which leads to (6.34) by postmultiplying by $e^{t\,Q}$ and taking into account that $\Pi(u) = e^{u\,Q}$, $u \geq 0$. □

If J is irreducible with stationary distribution $\pi = (\pi_j)$ we call the version of $(\mathbf{X}, J)$ for which $J(0)$ has distribution π the *stationary version* of the MAP of arrivals $(\mathbf{X}, J)$. We note that $\pi\Pi(t) = \pi$, for all t, and $\Pi(t) \to \mathbf{e}\pi$, as $t \to \infty$.

Corollary 6.1. *Suppose J is irreducible with stationary distribution $\pi = (\pi_j)$. For the stationary version of $(\mathbf{X}, J)$, and with $1 \leq i \leq r$, the following statements hold.*

(a) *Expected values of counts satisfy*

$$(E[X_i(t); J(t) = k]) = t\pi\overline{\Sigma}_1^{X_i}\mathbf{e}\pi + \pi\overline{\Sigma}_1^{X_i}\left(I - e^{tQ}\right)(\mathbf{e}\pi - Q)^{-1} \tag{6.43}$$

and, in particular,

$$E(X_i(t)) = t\pi\overline{\Sigma}_1^{X_i}\mathbf{e}. \tag{6.44}$$

(b) *Second order moments of the number of counts satisfy*

$$E(X_i^2(t)) = \left[t\pi\overline{\Sigma}_1^{X_i}\mathbf{e}\right]^2 + t\pi\overline{\Sigma}_2^{X_i}\mathbf{e} + 2t\pi\overline{\Sigma}_1^{X_i}C\overline{\Sigma}_1^{X_i}\mathbf{e}$$
$$-2\pi\overline{\Sigma}_1^{X_i}\left(I - e^{tQ}\right)(\mathbf{e}\pi - Q)^{-2}\overline{\Sigma}_1^{X_i}\mathbf{e} \tag{6.45}$$

with $C = (I - \mathbf{e}\pi)(\mathbf{e}\pi - Q)^{-1}$ and, in particular,

$$\frac{Var(X_i(t))}{t} \to \pi\left[\overline{\Sigma}_2^{X_i} + 2\overline{\Sigma}_1^{X_i}C\overline{\Sigma}_1^{X_i}\right]\mathbf{e}. \tag{6.46}$$

(c) *Covariances of counts satisfy*

$$\frac{Cov(X_i(t), X_l(t))}{t} \to c_{il} \tag{6.47}$$

where

$$c_{il} = \pi \left[\sum_{\mathbf{n} \in \mathbf{N}_+^r} n_i n_l \Lambda_{\mathbf{n}} + \overline{\Sigma}_1^{X_i} C \overline{\Sigma}_1^{X_l} + \overline{\Sigma}_1^{X_l} C \overline{\Sigma}_1^{X_i} \right] \mathbf{e}. \tag{6.48}$$

Proof. (a). In view of the fact that, since $\pi Q = \mathbf{0}$,

$$\left(e^{sQ} - \mathbf{e}\pi\right)\left(\mathbf{e}\pi - Q\right) = -e^{sQ}Q$$

we find that

$$\int_0^t \left(e^{sQ} - \mathbf{e}\pi\right) ds = \left(I - e^{tQ}\right)\left(\mathbf{e}\pi - Q\right)^{-1}. \tag{6.49}$$

As the previous equation implies that

$$\int_0^t e^{sQ}\, ds = \mathbf{e}\pi t + \left(I - e^{tQ}\right)\left(\mathbf{e}\pi - Q\right)^{-1}, \tag{6.50}$$

we obtain (6.43) by plugging (6.50) in (6.33). Moreover, (6.44) follows by postmultiplying both members of (6.43) by the vector $\mathbf{e}$ and using the fact that, in view of (6.49),

$$\left(I - e^{tQ}\right)\left(\mathbf{e}\pi - Q\right)^{-1}\mathbf{e} = \int_0^t \left(e^{sQ} - \mathbf{e}\pi\right)\mathbf{e}\, ds = \mathbf{0}.$$

(b). We first note that, in view of (6.34),

$$E(X_i^2(t)) = \pi \mathbb{E}_2^{X_i}(t)\,\mathbf{e} = 2\int_0^t \pi \mathbb{E}_1^{X_i}(s)\, ds\, \overline{\Sigma}_1^{X_i}\mathbf{e} + t\pi\overline{\Sigma}_2^{X_i}\mathbf{e}. \tag{6.51}$$

As, from (a), we have

$$\pi \mathbb{E}_1^{X_i}(s) = s\pi\overline{\Sigma}_1^{X_i}\mathbf{e}\pi + \pi\overline{\Sigma}_1^{X_i}\left(I - e^{sQ}\right)\left(\mathbf{e}\pi - Q\right)^{-1}$$

we get (6.45) from (6.51) after some algebraic manipulation taking into account (6.50). Moreover, (6.46) follows directly from (6.44) and (6.45).
(c). Proceeding as in Theorem 6.4 with $z_i = z_l = z$ and taking derivatives with respect to z, we would conclude that

$$\mathbb{E}_2^{X_i+X_l}(t) = \int_0^t \left[2\mathbb{E}_1^{X_i+X_l}(s)\overline{\Sigma}_1^{X_i+X_l} + \Pi(s)\overline{\Sigma}_2^{X_i+X_l}\right]\Pi(t-s)ds.$$

By the arguments used to prove (a) and (b) we may then conclude that

$$\frac{\mathrm{Var}(X_i(t) + X_l(t))}{t} \to \pi\left[\overline{\Sigma}_2^{X_i+X_l} + 2\left(\overline{\Sigma}_1^{X_i} + \overline{\Sigma}_1^{X_l}\right)C\left(\overline{\Sigma}_1^{X_i} + \overline{\Sigma}_1^{X_l}\right)\right]\mathbf{e}.$$

The statement follows easily from this and (6.46), using the fact that

$$\text{Cov}(X_i(t), X_l(t)) = \frac{1}{2}\left[\text{Var}(X_i + X_l(t)) - \text{Var}(X_i(t)) - \text{Var}(X_l(t))\right],$$

$\overline{\Sigma}_1^{X_i+X_l} = \overline{\Sigma}_1^{X_i} + \overline{\Sigma}_1^{X_l}$, and

$$\overline{\Sigma}_2^{X_i+X_l} = \overline{\Sigma}_2^{X_i} + 2\sum_{\mathbf{n}\in\mathbf{N}_+^r} n_i n_l \Lambda_{\mathbf{n}} + \overline{\Sigma}_2^{X_l}.$$

□

In case J has a stationary distribution $\pi = (\pi_j)$, we let $\lambda = (\lambda_1, \ldots, \lambda_r)$, with

$$\lambda_i = \pi\overline{\Sigma}_1^{X_i}\mathbf{e} \quad (1 \le i \le r). \tag{6.52}$$

As the next law of large numbers shows, λ denotes the vector of asymptotic individual arrival rates associated to the coordinates of $\mathbf{X}$.

Theorem 6.5. (Law of Large Numbers).
Suppose J is irreducible and has stationary distribution π, and $\lambda_i < \infty, \forall i$, then for all initial distributions

$$\frac{\mathbf{X}(t)}{t} \to \lambda \qquad a.s. \tag{6.53}$$

Proof. Consider the sequence $\{T_p\}_{p\ge 0}$ of successive transition epochs in $(\mathbf{X}, J)$, with $T_0 = 0$. We assume w.l.o.g. that $T_1 < \infty$ a.s., since otherwise $\mathbf{X}(t) \equiv \mathbf{0}$ a.s., $\Lambda_{\mathbf{n}} = 0$, $\mathbf{n} \in \mathbf{N}_+^r$, and everything is satisfied trivially. If for fixed i, $1 \le i \le r$, we define the process $\{(T_p, Y_p, J_p), p \ge 0\}$, with $J_p = J(T_p)$ and $Y_p = X_i(T_p)$, then

$$(E_j(Y_1)) = (E(Y_1 \mid J_0 = j)) = \left(\frac{1}{\gamma_j}\sum_{k\in E}\sum_{\mathbf{n}\in\mathbf{N}_+^r} n_i\lambda_{jk}(\mathbf{n})\right) = \Gamma^{-1}\overline{\Sigma}_1^{X_i}\mathbf{e}$$

and $(E_j(T_1)) = (E(T_1 \mid J_0 = j)) = (1/\gamma_j) = \Gamma^{-1}\mathbf{e}$. Moreover $\{J_p\}$ is an irreducible discrete time Markov chain with stationary distribution $\pi^\star = \beta^{-1}\pi\Gamma$, where $\beta = \sum \pi_k\gamma_k$. This implies by the ergodic theory for Markov chains that a.s.

$$\frac{Y_n}{n} \to \sum_{j\in E}\pi_j^\star E_j(Y_1) = \beta^{-1}\pi\overline{\Sigma}_1^{X_i}\mathbf{e} \tag{6.54}$$

and

$$\frac{T_n}{n} \to \sum_{j\in E}\pi_j^\star E_j(T_1) = \beta^{-1} \tag{6.55}$$

for any initial distribution. For $t \in \mathbf{R}_+$, we let

$$n(t) = \max\{p \in \mathbf{N}_+ : T_p \leq t\}.$$

Note that, by (6.10), $n(t) < \infty$ a.s., and, moreover, $n(t) \uparrow \infty$ a.s. as $t \uparrow \infty$. Now, since

$$\begin{aligned} \frac{Y_{n(t)}}{n(t)} \frac{n(t)}{n(t)+1} \frac{n(t)+1}{T_{n(t)+1}} = \frac{Y_{n(t)}}{T_{n(t)+1}} &\leq \frac{X_i(t)}{t} \\ &\leq \frac{Y_{n(t)+1}}{T_{n(t)}} = \frac{Y_{n(t)+1}}{n(t)+1} \frac{n(t)+1}{n(t)} \frac{n(t)}{T_{n(t)}}, \end{aligned}$$

we conclude, using (6.54) and (6.55), that $\frac{X_i(t)}{t} \to \pi \overline{\Sigma}_1^{X_i} \mathbf{e}$ a.s., and (6.53) follows. □

We note that for the stationary version of $(\mathbf{X}, J)$, (6.53) may be obtained from (6.44) and (6.46) by using Chebyshev's inequality, as in the proof of the law of large numbers for the Poisson process in [56]. Moreover, a central limit theorem for the additive component X of an MAP of arrivals $(\mathbf{X}, J)$ can be derived in case the Markov component J is finite and positive recurrent using, e.g., [52]. We state the result without proof.

Theorem 6.6. *Suppose $(\mathbf{X}, J)$ is an MAP of arrivals on $\mathbf{N}_+^r \times E$ such that J is finite and irreducible with stationary distribution π and $\pi \overline{\Sigma}_2^{X_i} \mathbf{e} < \infty \, (1 \leq i \leq r)$. Then*

$$\sqrt{t}\left(\frac{\mathbf{X}(t)}{t} - \lambda\right) \stackrel{\mathcal{L}}{\Longrightarrow} \mathbf{N}(\mathbf{0}, C) \tag{6.56}$$

where $\stackrel{\mathcal{L}}{\Longrightarrow}$ denotes convergence in distribution and $\mathbf{N}(\mathbf{0}, C)$ denotes a multivariate normal distribution with null mean vector and covariance matrix $C = (c_{il})$, with c_{il} given by (6.48).

6.5 Some Properties of MAPs

In this section we study some properties of MAPs, with emphasis on transformations that preserve the Markov-additive property. We will use the results derived in this section in the next two sections, where we address transformations of MAPs of arrivals.

For clarity, we consider only time-homogeneous MAPs, but note that some of the results may be easily adapted to the non-homogeneous case.

Moreover, we note that the distribution of an MAP $(\mathbf{X}, J)$ on $\mathbf{R}^r \times E$ is characterized by its transition probability measure

$$P_{jk}(A;t) = P\{\mathbf{X}(t) \in A,\ J(t) = k \mid J(0) = j\} \tag{6.57}$$

for $t \geq 0$, $j, k \in E$, and a Borel subset A of $\mathbf{R}^r$. In the following, we let $\mathcal{R}^r$ denote the Borel σ-algebra on $\mathbf{R}^r$.

We start with the following result which presents important properties of deterministic transformations of MAPs.

Theorem 6.7. *Let $A \in \mathcal{R}^r$ and $B \in \mathcal{R}^s$.*
(a). (**Linear transformations of MAPs**).
Suppose that $(\mathbf{X}, J)$ is an MAP on $\mathbf{R}^r \times E$ with transition probability measure $P^{\mathbf{X}}$, and $T : \mathbf{R}^r \to \mathbf{R}^s$ is a linear transformation.

If $\mathbf{Y} = T(\mathbf{X})$, then $(\mathbf{Y}, J)$ is an MAP on $\mathbf{R}^s \times E$ with transition probability measure $P^{\mathbf{Y}}$ such that

$$P^{\mathbf{Y}}_{jk}(B;t) = P^{\mathbf{X}}_{jk}\left(T^{-1}(B);t\right). \tag{6.58}$$

(b). (**Patching together independent MAPs**).
Suppose that $(\mathbf{X}, J_1)$ and $(\mathbf{Y}, J_2)$ are MAPs on $\mathbf{R}^r \times E_1$ and $\mathbf{R}^s \times E_2$ with transition probability measures $P^{\mathbf{X}}$ and $P^{\mathbf{Y}}$, respectively.

If $(\mathbf{X}, J_1)$ and $(\mathbf{Y}, J_2)$ are independent, then $((\mathbf{X}, \mathbf{Y}), (J_1, J_2))$ is an MAP on $\mathbf{R}^{r+s} \times E_1 \times E_2$ with transition probability measure $P^{(\mathbf{X},\mathbf{Y})}$ such that

$$P^{(\mathbf{X},\mathbf{Y})}_{(j_1,j_2)(k_1,k_2)}(A \times B;t) = P^{\mathbf{X}}_{j_1k_1}(A;t)\, P^{\mathbf{Y}}_{j_2k_2}(B;t). \tag{6.59}$$

Proof. (a). Theorem II(1.2) of [17] states sufficient conditions under which transformations of Markov processes are themselves Markov. These conditions do not apply necessarily to our case, but we may prove the desired result more directly as follows.

Suppose $0 = t_0 < t_1 < \ldots < t_n < t$, $\mathbf{y}_p \in \mathbf{R}^s$ $(0 \leq p \leq n)$, $B \in \mathcal{R}^s$ and let $A_p = \{(\mathbf{Y}(t_p), J(t_p)) = (\mathbf{y}_p, j_p)\}$, $0 \leq p \leq n$. From the fact that T is a measurable transformation we have

$$A_p = \{(\mathbf{X}(t_p), J(t_p)) \in T^{-1}(\{\mathbf{y}_p\}) \times \{j_p\}\} \in \sigma(\mathbf{X}(t_p), J(t_p)),$$

for $0 \leq p \leq n$, and, similarly, $A_t = \{(\mathbf{X}(t), J(t)) \in T^{-1}(B) \times \{k\}\}$ belongs to $\sigma((\mathbf{X}, J)(t))$, where $\sigma(\mathbf{X}(s), J(s))$ denotes the σ-algebra generated by

$(\mathbf{X}(s), J(s))$. Since $(\mathbf{X}, J)$ is an MAP, these facts imply that

$$\begin{aligned}
P\{\mathbf{Y}(t) &\in B;\ J(t) = k \mid \cap_{0\leq p\leq n} A_p\} \\
&= P\{\mathbf{Y}(t) \in B;\ J(t) = k \mid \mathbf{Y}(t_n) = \mathbf{y}_n,\ J(t_n) = j_n\} \\
&= P\{\mathbf{Y}(t) - \mathbf{Y}(t_n) \in B - \mathbf{y}_n;\ J(t) = k \mid A_n\} \\
&= P\{\mathbf{X}(t) - \mathbf{X}(t_n) \in T^{-1}(B - \mathbf{y}_n);\ J(t) = k \mid J(t_n) = j_n\} \\
&= P\{\mathbf{X}(t - t_n) \in T^{-1}(B - \mathbf{y}_n);\ J(t - t_n) = k \mid J(0) = j_n\} \\
&= P\{\mathbf{Y}(t - t_n) \in B - \mathbf{y}_n;\ J(t - t_n) = k \mid J(0) = j_n\}.
\end{aligned}$$

This shows that $(\mathbf{Y}, J)$ is an MAP with the given transition probability measure.

(b). Since $(\mathbf{X}, J_1)$ and $(\mathbf{Y}, J_2)$ are independent time-homogeneous Markov processes, it is clear that $(\mathbf{X}, \mathbf{Y}, J_1, J_2)$ is a time-homogeneous Markov process. Property (b) of MAPs for $((\mathbf{X}, \mathbf{Y}), (J_1, J_2))$ follows using arguments similar to the ones needed to prove that $(\mathbf{X}, \mathbf{Y}, J_1, J_2)$ is Markov, and its proof is ommited. □

We note that, in Theorem 6.7, if the original processes are strong MAPs so are the transformed processes. In addition, if the original processes are Markov subordinators so are the resulting processes obtained from patching together independent processes or by taking linear transformations with nonnegative coefficients. Similar remarks apply to Corollary 6.2 bellow.

Corollary 6.2. *Suppose* $\alpha, \beta \in \mathbf{R}$, $A \in \mathcal{R}^r$, *and* $B \in \mathcal{R}^s$.

(a). (**Marginals of MAPs**).

If $(\mathbf{X}, J)$ *is an MAP on* $\mathbf{R}^r \times E$ *with transition probability measure* $P^{\mathbf{X}}$ *and*

$$\mathbf{Z} = (X_{i_1}, X_{i_2}, \ldots, X_{i_s}), \quad 1 \leq i_1 < \ldots < i_s \leq r$$

then $(\mathbf{Z}, J)$ *is an MAP on* $\mathbf{R}^s \times E$. *If, without loss of generality,* $\mathbf{Z} = (X_1, X_2, \ldots, X_s)$, *then* $(\mathbf{Z}, J)$ *has transition probability measure*

$$P_{jk}(B) = P_{jk}^{\mathbf{X}}(B \times \mathbf{R}^{r-s}). \tag{6.60}$$

(b). (**Linear combinations of dependent MAPs**).

If $\mathbf{X} = (X_1, \ldots, X_r)$, $\mathbf{Y} = (Y_1, \ldots, Y_r)$ *and* $((\mathbf{X}, \mathbf{Y}), J)$ *is an MAP on* $\mathbf{R}^{2r} \times E$ *with transition probability measure* $P^{(\mathbf{X},\mathbf{Y})}$, *then the process* $(\alpha\mathbf{X} + \beta\mathbf{Y}, J)$ *is an MAP on* $\mathbf{R}^r \times E$ *with transition probability measure*

$$P_{jk}(A; t) = P_{jk}^{(\mathbf{X},\mathbf{Y})}\left(\{\mathbf{z} = (\mathbf{x}, \mathbf{y}),\ \mathbf{x}, \mathbf{y} \in \mathbf{R}^r\ :\ \alpha\mathbf{x} + \beta\mathbf{y} \in A\}\,; t\right). \tag{6.61}$$

(c). (**Linear combinations of independent MAPs**).

If $(\mathbf{X}, J_1)$ *and* $(\mathbf{Y}, J_2)$ *are independent MAPs on* $\mathbf{R}^r \times E_1$ *and* $\mathbf{R}^r \times E_2$

with transition probability measure $P^{\mathbf{X}}$ *and* $P^{\mathbf{Y}}$*, respectively, then* $(\alpha\mathbf{X} + \beta\mathbf{Y}, (J_1, J_2))$ *is an MAP on* $\mathbf{R}^r \times E_1 \times E_2$ *with transition probability measure*

$$P_{(j_1,j_2)(k_1,k_2)}(A;t) = \int_{A(\alpha,\beta)} P^{\mathbf{X}}_{j_1k_1}(d\mathbf{x};t)\, P^{\mathbf{Y}}_{j_2k_2}(d\mathbf{y};t), \tag{6.62}$$

where $A(\alpha,\beta) = \{\mathbf{z} = (\mathbf{x},\mathbf{y}),\, \mathbf{x},\mathbf{y} \in \mathbf{R}^r \,:\, \alpha\mathbf{x} + \beta\mathbf{y} \in A\}$.

Proof. Statements (a) and (b) follow from Theorem 6.7 (a) since marginals and linear combinations are special cases of linear transformations. To prove statement (c), we first patch together $(\mathbf{X}, J_1)$ and $(\mathbf{Y}, J_2)$ to get in their product probability space the process $((\mathbf{X},\mathbf{Y}),(J_1,J_2))$, which by Theorem 6.7 (b) is an MAP with probability transition measure given by (6.59). Using (b), it then follows that $(\alpha\mathbf{X} + \beta\mathbf{Y}, (J_1, J_2))$ is an MAP of arrivals with transition probability measure as given. □

If $(\mathbf{X}, J)$ is an MAP on $\mathbf{R}^r \times E$ and $0 = t_0 \leq t_1 \leq t_2 \leq \ldots$ is a deterministic sequence, then the process $(S^\star, J^\star) = \{(S^\star_n, J^\star_n) = (\mathbf{X}(t_n), J(t_n)),\, n \in \mathbf{N}_+\}$ is an MRW on $\mathbf{R}^r \times E$ (which is time-homogeneous in case $t_n = nh, n \in \mathbf{N}_+$, for some $h > 0$). In order for the imbedded process to be an MRW, the imbedded sequence does not need to be deterministic, and more interesting imbeddings may be obtained by using a random sequence of imbedding times.

For the next result, we recall that for a time-homogeneous MRW $(\mathbf{S}^\star, J^\star)$ on $\mathbf{R}^r \times E$, it suffices to define the one-step transition probability measure

$$V_{jk}(A) = P\{\mathbf{S}^\star_1 \in A;\, J^\star_1 = k \mid J^\star_0 = j\}$$

for $j,k \in E$ and $A \in \mathcal{R}^r$, along with the initial distribution of $J^\star$.

Theorem 6.8. *Suppose* $(\mathbf{X}, J)$ *is an MAP on* $\mathbf{R}^r \times E$*, and* $(T^\star_n)_{n\geq 0}$ *are stopping times such that* $0 = T^\star_0 \leq T^\star_1 \leq \ldots < \infty$ *a.s. Denote* $J^\star_n = J(T^\star_n)$*, and* $\mathbf{S}^\star_n = \mathbf{X}(T^\star_n)$*, for* $n \in \mathbf{N}_+$.

If $(\mathbf{X}, J)$ *is a strong MAP, then* $(T^\star, \mathbf{S}^\star, J^\star) = \{(T^\star_n, \mathbf{S}^\star_n, J^\star_n),\, n \in \mathbf{N}_+\}$ *is an MRW on* $\mathbf{R}_+ \times \mathbf{R}^r \times E$*. Moreover, in case the conditional distribution of* $(T^\star_{n+1} - T^\star_n, S^\star_{n+1} - S^\star_n, J^\star_{n+1})$ *given* $J^\star_n$ *does not depend on* n*, then* $(T^\star, \mathbf{S}^\star, J^\star)$ *is an homogeneous MRW with one-step transition probability measure*

$$V_{jk}(A \times B) = P\{T^\star_1 \in A;\, \mathbf{X}(T^\star_1) \in B;\, J(T^\star_1) = k \mid J(0) = j\}$$

for $j,k \in E$*,* $B \in \mathcal{R}^r$*, and a Borel set* A *on* $\mathbf{R}_+$.

In applications, the most common uses of Theorem 6.8 are for the case where the imbedding points are the successive transition epochs in J and the case where the sequence $T^\star$ is a renewal sequence independent of $(\mathbf{X}, J)$ as addressed in [65].

6.6 Transformations of MAPs of Arrivals

In this section we consider transformations of MAPs of arrivals. We study first deterministic transformations and later random transformations.

Considering transformations of MAPs, and for $\mathbf{Z} = D\mathbf{X}$ with $D_{p\times r}$ being a matrix of constants with values in $\mathbf{N}_+$, it is useful to define for $\mathbf{m} > \mathbf{0}$

$$\Lambda_{\mathbf{m}}^{\mathbf{Z}} = \sum_{\{\mathbf{n}\in\mathbf{N}_+^r : D\mathbf{n}=\mathbf{m}\}} \Lambda_{\mathbf{n}}. \tag{6.63}$$

Theorem 6.9.
(a). **(Linear transformations of MAPs of arrivals).**
If $(\mathbf{X}, J)$ is an MAP of arrivals on $\mathbf{N}_+^r \times E$ with $(Q, \{\Lambda_{\mathbf{n}}\}_{\mathbf{n}>\mathbf{0}})$-source and $\mathbf{Z} = D\mathbf{X}$ where $D_{p\times r}$ is a matrix of constants with values in $\mathbf{N}_+$, then $(\mathbf{Z}, J)$ is an MAP of arrivals on $\mathbf{N}_+^p \times E$ with $(Q, \{\Lambda_{\mathbf{m}}^{\mathbf{Z}}\}_{\mathbf{m}>\mathbf{0}})$-source.
(b). **(Patching together independent MAPs of arrivals).**
Suppose $(\mathbf{X}, J_1)$ and $(\mathbf{Y}, J_2)$ are independent MAPs of arrivals on $\mathbf{N}_+^r \times E_1$ and $\mathbf{N}_+^s \times E_2$ with sources $(Q^1, \{\Lambda_{\mathbf{n}}^1\}_{\mathbf{n}>\mathbf{0}})$ and $(Q^2, \{\Lambda_{\mathbf{m}}^2\}_{\mathbf{m}>\mathbf{0}})$, respectively. If $(\mathbf{X}, J_1)$ and $(\mathbf{Y}, J_2)$ are independent, then $((\mathbf{X}, \mathbf{Y}), (J_1, J_2))$ is an MAP of arrivals on $\mathbf{N}_+^{r+s} \times E_1 \times E_2$ with $(Q^1 \oplus Q^2, \{\Lambda_{(\mathbf{n},\mathbf{m})}\}_{(\mathbf{n},\mathbf{m})>\mathbf{0}})$-source, where

$$\Lambda_{(\mathbf{n},\mathbf{m})} = [\Lambda_{\mathbf{n}}^1 \mathbf{1}_{\{\mathbf{m}=\mathbf{0}\}}] \oplus [\mathbf{1}_{\{\mathbf{n}=\mathbf{0}\}} \Lambda_{\mathbf{m}}^2] \tag{6.64}$$

for $(\mathbf{n}, \mathbf{m}) > \mathbf{0}$.

Proof. (a). The fact that $(\mathbf{Z}, J)$ is an MAP follows from Theorem 6.7 (a), and since D has entries with values in $\mathbf{N}_+$ it follows that $\mathbf{Z}$ takes values in $\mathbf{N}_+^p$. Thus $(\mathbf{Z}, J)$ is an MAP of arrivals on $\mathbf{N}_+^p \times E$; moreover, using (6.58) and (6.5), we have for $(j, k, \mathbf{m}) \in S_p(E)$

$$\begin{aligned} P(\mathbf{Z}(h) = \mathbf{m}, J(h) = k \mid J(0) = j) &= \sum_{\{\mathbf{n}\in\mathbf{N}_+^r : D\mathbf{n}=\mathbf{m}\}} P(\mathbf{X}(h) = \mathbf{n}, J(h) = k \mid J(0) = j) \\ &= h \sum_{\{\mathbf{n}\in\mathbf{N}_+^r : D\mathbf{n}=\mathbf{m}\}} \lambda_{jk}(\mathbf{n}) + o(h). \end{aligned}$$

This implies that $(\mathbf{Z}, J)$ has $(Q, \{\Lambda^{\mathbf{Z}}_{\mathbf{m}}\}_{\mathbf{m}>\mathbf{0}})$-source.
(b). Using Theorem 6.7 (b), it follows easily that $((\mathbf{X}, \mathbf{Y}), (J_1, J_2))$ is an MAP of arrivals on $\mathbf{N}_+^{r+s} \times E_1 \times E_2$. From the independence of J_1 and J_2, it is well known (and easy to check) that (J_1, J_2) has generator matrix $Q^1 \oplus Q^2$. We let P^1 and P^2 be the transition probability measures of $(\mathbf{X}, J_1)$ and $(\mathbf{Y}, J_2)$, respectively. For $((j_1, j_2), (k_1, k_2), (\mathbf{n}, \mathbf{m})) \in S_{r+s}(E_1 \times E_2)$ the transition probability measure of $((\mathbf{X}, \mathbf{Y}), (J_1, J_2))$ is such that

$$
\begin{aligned}
&\left(P_{(j_1,j_2)(k_1,k_2)}((\mathbf{n}, \mathbf{m}); h)\right) \\
&\quad = \left(\lambda^{\mathbf{1}}_{j_1 k_1}(\mathbf{n}) \mathbf{1}_{\{\mathbf{m}=\mathbf{0}\}} \delta_{j_2 k_2} + \delta_{j_1 k_1} \mathbf{1}_{\{\mathbf{n}=\mathbf{0}\}} \lambda^{2}_{j_2 k_2}(\mathbf{m})\right) h + o(h) \\
&\quad = \left([\Lambda^{\mathbf{1}}_{\mathbf{n}} \mathbf{1}_{\{\mathbf{m}=\mathbf{0}\}}] \oplus [\mathbf{1}_{\{\mathbf{n}=\mathbf{0}\}} \Lambda^{2}_{\mathbf{m}}]\right) h + o(h)
\end{aligned}
$$

which proves the statement. □

Some consequences of Theorem 6.9 are stated in Corollary 6.3, which is stated without proof. The statements in Corollary 6.3 may be proved in a way similar to that used to prove the consequences of Theorem 6.7 which constitute Corollary 6.2.

Corollary 6.3. *Assume that* $\alpha, \beta \in \mathbf{N}_+$.
(a). (**Marginals of MAPs of arrivals**).
If $(\mathbf{X}, J)$ *is an MAP of arrivals on* $\mathbf{N}_+^r \times E$ *with* $(Q, \{\Lambda_{\mathbf{n}}\}_{\mathbf{n}>\mathbf{0}})$*-source and* $\mathbf{Z} = (X_{i_1}, \ldots, X_{i_p})$ *with* $1 \leq i_1 < \ldots < i_p \leq r$*, then* $(\mathbf{Z}, J)$ *is an MAP of arrivals on* $\mathbf{N}_+^p \times E$ *with* $(Q, \{\Lambda^{\mathbf{Z}}_{\mathbf{m}}\}_{\mathbf{m}>\mathbf{0}})$*-source, where for* $\mathbf{m} > \mathbf{0}$

$$
\Lambda^{\mathbf{Z}}_{\mathbf{m}} = \sum_{\{\mathbf{n}:(n_{i_1}, n_{i_2}, \ldots, n_{i_p})=\mathbf{m}\}} \Lambda_{\mathbf{n}}. \tag{6.65}
$$

(b). (**Linear combinations of dependent MAPs of arrivals**).
If $\mathbf{X} = (X_1, \ldots, X_r)$, $\mathbf{Y} = (Y_1, \ldots, Y_r)$*, and* $((\mathbf{X}, \mathbf{Y}), J)$ *is an MAP of arrivals on* $\mathbf{N}_+^{2r} \times E$ *with* $(Q, \{\Lambda_{(\mathbf{n},\mathbf{m})}\}_{(\mathbf{n},\mathbf{m})>\mathbf{0}})$*-source, then the process* $(\mathbf{Z}, J) = (\alpha\mathbf{X}+\beta\mathbf{Y}, J)$ *is an MAP of arrivals on* $\mathbf{N}_+^r \times E$ *with* $(Q, \{\Lambda^{\mathbf{Z}}_{\mathbf{a}}\}_{\mathbf{a}>\mathbf{0}})$*-source, where for* $\mathbf{a} > \mathbf{0}$

$$
\Lambda^{\mathbf{Z}}_{\mathbf{a}} = \sum_{\{\mathbf{n},\mathbf{m} \in \mathbf{N}_+^r : \alpha\mathbf{n}+\beta\mathbf{m}=\mathbf{a}\}} \Lambda_{(\mathbf{n},\mathbf{m})}. \tag{6.66}
$$

(c). (**Linear combinations of independent MAPs of arrivals**).
Suppose $(\mathbf{X}, J_1)$ *and* $(\mathbf{Y}, J_2)$ *are independent MAPs of arrivals on* $\mathbf{N}_+^r \times E_1$ *and* $\mathbf{N}_+^r \times E_2$ *with sources* $(Q^1, \{\Lambda^{\mathbf{1}}_{\mathbf{n}}\}_{\mathbf{n}>\mathbf{0}})$ *and* $(Q^2, \{\Lambda^{2}_{\mathbf{m}}\}_{\mathbf{m}>\mathbf{0}})$*, respectively. If* $\mathbf{Z} = \alpha\mathbf{X} + \beta\mathbf{Y}$ *and* $J = (J_1, J_2)$*, then* $(\mathbf{Z}, J)$ *is an MAP of arrivals on* $\mathbf{N}_+^r \times E_1 \times E_2$ *with* $(Q^1 \oplus Q^2, \{\Lambda^{\mathbf{Z}}_{\mathbf{a}}\}_{\mathbf{a}>\mathbf{0}})$*-source. Here*

$$
\Lambda^{\mathbf{Z}}_{\mathbf{a}} = \sum_{\{\mathbf{n},\mathbf{m} \in \mathbf{N}_+^r : \alpha\mathbf{n}+\beta\mathbf{m}=\mathbf{a}\}} \Lambda_{(\mathbf{n},\mathbf{m})} \tag{6.67}
$$

for $\mathbf{a} > \mathbf{0}$, *where* $\Lambda_{(\mathbf{n},\mathbf{m})}$ *is given by (6.64).*

Theorem 6.9 and Corollary 6.3 show that the class of MAPs of arrivals is closed under important transformations. Moreover the transition rates of the transformed processes are easily obtained. We note that Corollary 6.3 (b) and (c) could have been stated for linear combinations with a finite number of terms. Corollary 6.3 (c) leads to the following result which has been used to establish certain asymptotic results and to study the effect of multiplexing bursty traffic streams in an ATM network (see [37] and [63] for references).

Corollary 6.4 (Finite sums of independent MAPs of arrivals). *If* $(\mathbf{X}_1, J_1), \ldots, (\mathbf{X}_K, J_K)$ *are independent MAPs of arrivals on* $\mathbf{N}_+^r \times E_1, \ldots, \mathbf{N}_+^r \times E_K$ *with sources* $(Q^1, \{\Lambda_\mathbf{n}^1\}_{\mathbf{n}>\mathbf{0}}), \ldots, (Q^K, \{\Lambda_\mathbf{n}^K\}_{\mathbf{n}>\mathbf{0}})$, *then*

$$(\mathbf{X}_1 + \mathbf{X}_2 + \ldots + \mathbf{X}_K, (J_1, J_2, \ldots, J_K))$$

is an MAP of arrivals on $\mathbf{N}_+^r \times E_1 \times E_2 \times \ldots E_K$ *with source*

$$\left(Q^1 \oplus Q^2 \oplus \ldots \oplus Q^K, \{\Lambda_\mathbf{n}^1 \oplus \Lambda_\mathbf{n}^2 \oplus \ldots \oplus \Lambda_\mathbf{n}^K\}_{\mathbf{n}>\mathbf{0}}\right). \tag{6.68}$$

Proof. It suffices to prove the statement for $K = 2$. Using Corollary 6.3 (c) we conclude that $(\mathbf{X}_1 + \mathbf{X}_2, (J_1, J_2))$ is an MAP of arrivals on $\mathbf{N}_+^r \times E_1 \times E_2$ with $(Q_1 \oplus Q_2, \{\Lambda_\mathbf{n}^\star\}_{\mathbf{n}>\mathbf{0}})$ where for $\mathbf{n} > \mathbf{0}$

$$\Lambda_\mathbf{n}^\star = \Lambda_\mathbf{n}^1 \oplus 0 + 0 \oplus \Lambda_\mathbf{n}^2 = \Lambda_\mathbf{n}^1 \oplus \Lambda_\mathbf{n}^2.$$

□

Corollary 6.4 shows that the class of MAPs of arrivals is closed under finite superpositions of independent processes, which is a generalization of the similar result for Poisson processes and for MMPPs. The proof for the MMPP in [72] is based essentially in properties of the Poisson process (this is because the MMPP process may be viewed as constructed from a series of independent Poisson processes such that the i-th Poisson process is observed only when the Markov component is in state i, as remarked in Sec. 6.2). For more general MAPs of arrivals, it is better to base the correspondent proof more directly on the Markov property (as in the proof given above), which is also the basic property of the Poisson process leading to the additive property.

In computational terms, the rapid geometric increase in the dimensionality of the state space of the Markov component of

$$(\mathbf{X}_1 + \mathbf{X}_2 + \ldots + \mathbf{X}_K, (J_1, J_2, \ldots, J_K))$$

as the number K of added independent MAPs of arrivals increases, limits the utility of Corollary 6.4. We show in Example 6.4 that this unpleasant situation may be avoided in the case where the processes added have identical parameters.

Example 6.4. (Superposition of independent and identical MAPs of arrivals).

Suppose that the processes $(\mathbf{X}_i, J_i)$ $(1 \leq i \leq K)$ are independent MAPs of arrivals on $\mathbf{N}_+^r \times \{0, 1, \ldots, m\}$ with common rate matrices $\Lambda_{\mathbf{n}}$. If we define

$$J_p^\star(t) = \#\{1 \leq i \leq K : J_i(t) = p\} \qquad (1 \leq p \leq m)$$

then $(J_1^\star, J_2^\star, \ldots, J_m^\star)$ is a Markov process on the state space

$$E = \{(i_1, i_2, \ldots, i_m) \in \mathbf{N}_+^m : 0 \leq i_1 + i_2 + \ldots + i_m \leq K\}.$$

Moreover, it may be seen that the process, $(\mathbf{X}_1 + \mathbf{X}_2 + \ldots + \mathbf{X}_K, (J_1^\star, J_2^\star, \ldots, J_m^\star))$ is an MAP of arrivals on $\mathbf{N}_+^r \times E$ whose transition rates $\lambda_{\mathbf{ij}}^\star(\mathbf{n})$ are such that for $(\mathbf{i}, \mathbf{j}, \mathbf{n}) \in S_r(E)$

$$\lambda_{\mathbf{ij}}^\star(\mathbf{n}) = \begin{cases} 0 & \#d(\mathbf{i}, \mathbf{j}) > 2 \\ i_l \lambda_{lp}(\mathbf{n}) & i_l - j_l = j_p - i_p = 1, d(\mathbf{i}, \mathbf{j}) = \{l, p\} \\ \sum_{l=0}^m i_l \lambda_{ll}(\mathbf{n}) & \mathbf{i} = \mathbf{j} \end{cases}$$

where

$$i_0 = K - \sum_{l=1}^m i_l, \qquad j_0 = K - \sum_{l=1}^m j_l, \qquad d(\mathbf{i}, \mathbf{j}) = \{l : 0 \leq l \leq m, i_l \neq j_l\}$$

Particular cases of this example have been considered by some authors (see e.g. [37] and [72] for superpositioning of two-state MMPPs, which are also called *switched Poisson processes*).

In Example 6.4 the reduction in the number of states from $(J_1, \ldots, J_K)$ to $(J_1^\star, \ldots, J_m^\star)$ increases with the number of added processes K and is specially significant when K is large compared with m. In the special case where the individual MAPs of arrivals have a two-state Markov component $(m = 1)$ the state space is reduced from 2^K states to $K + 1$ states. This result is important in applications where MAPs of arrivals with a two-state Markov component have been used to model the input from bursty sources in communications systems (see [37] and [63] for references).

Theorem 6.10. (Random time transformations of MAPs of arrivals).
Suppose that $(\mathbf{X}, J)$ *is an MAP of arrivals on* $\mathbf{N}_+^r \times E$ *with* $(Q, \{\Lambda_{\mathbf{n}}\}_{\mathbf{n}>\mathbf{0}})$*-source,* f *is a nonnegative function on* E*,* $A = (a_{jk}) = (f(j)\delta_{jk})$*, and*

$$A(t) = \int_0^t f(J(s))\, ds. \tag{6.69}$$

(a). *With* $(\mathbf{X}^A(t), J^A(t)) = (\mathbf{X}(A(t)), J(A(t)))$*, the process* $(\mathbf{X}^A, J^A)$ *is an MAP of arrivals on* $\mathbf{N}_+^r \times E$ *with* $(AQ, \{A\Lambda_{\mathbf{n}}\}_{\mathbf{n}>\mathbf{0}})$*-source.*
(b). *Suppose* $\inf_{j\in E} f(j) > 0$*, and for* $t \in \mathbf{R}_+$ *let*

$$B(t) = \inf\{x > 0 : A(x) \geq t\}. \tag{6.70}$$

With $(\mathbf{X}^B(t), J^B(t)) = (\mathbf{X}(B(t)), J(B(t)))$*, the process* $(\mathbf{X}^B, J^B)$ *is an MAP of arrivals on* $\mathbf{N}_+^r \times E$ *with* $(A^{-1}Q, \{A^{-1}\Lambda_{\mathbf{n}}\}_{\mathbf{n}>\mathbf{0}})$*-source.*

Proof. (a). Since $A(t)$ belongs to the σ-algebra generated by $\{(\mathbf{X}, J)(s), 0 \leq s \leq t\}$ and (A, J) is a time-homogeneous Markov subordinator, properties (a) and (b) of MAPs for $(\mathbf{X}^A, J^A)$ follow directly from the correspondent properties for the MAP $(\mathbf{X}, J)$, and moreover $(\mathbf{X}^A, J^A)$ is time-homogeneous. Thus $(\mathbf{X}^A, J^A)$ is a time-homogeneous MAP with the same state space as $(\mathbf{X}, J)$. We denote by P^A the transition probability measure of $(\mathbf{X}^A, J^A)$. For $(j, k, \mathbf{n}) \in S_r(E)$ we have

$$\begin{aligned} P_{jk}^A(\mathbf{n}; h) &= \int_0^{f(j)h} e^{-\gamma_j u}\lambda_{jk}(\mathbf{n})e^{-\gamma_k f(k)(h-u)}du + o(h) \\ &= f(j)\lambda_{jk}(\mathbf{n})h + o(h), \end{aligned}$$

from which we conclude that $(\mathbf{X}^A, J^A)$ has $(AQ, \{A\Lambda_{\mathbf{n}}\}_{\mathbf{n}>\mathbf{0}})$-source.
(b). First note that since $B(t) \leq x \Leftrightarrow A(x) \geq t$, it follows that $B(t)$ is a stopping time for J. Using the fact that $(\mathbf{X}, J)$ is a strong MAP and (A, J) is a time-homogeneous Markov subordinator, this implies that properties (a) and (b) of MAPs for $(\mathbf{X}^B, J^B)$ follow directly from the correspondent properties for $(\mathbf{X}, J)$, and that $(\mathbf{X}^B, J^B)$ is a time-homogeneous process. Thus $(\mathbf{X}^B, J^B)$ is a time-homogeneous MAP with the same state space as $(\mathbf{X}, J)$. We denote by P^B the transition probability measure of $(\mathbf{X}^B, J^B)$. For $(j, k, \mathbf{n}) \in S_r(E)$ we have

$$\begin{aligned} P_{jk}^B(\mathbf{n}; h) &= \int_0^{[f(j)]^{-1}h} e^{-\gamma_j u}\lambda_{jk}(\mathbf{n})e^{-\gamma_k \frac{h-f(j)u}{f(k)}}du + o(h) \\ &= [f(j)]^{-1}\lambda_{jk}(\mathbf{n})h + o(h), \end{aligned}$$

so that $(\mathbf{X}^B, J^B)$ has $(A^{-1}Q, \{A^{-1}\Lambda_{\mathbf{n}}\}_{\mathbf{n}>\mathbf{0}})$-source. □

The functional A used in Theorem 6.10 is a continuous additive functional of the Markov component of (X, J) (for definition see e.g. [17] and [28]). Functionals of the same type as A in (6.69) have have been considered in the applied literature (e.g. [73], and [93]); they may be used in particular to model the service in queueing systems with variable service rate and will be considered in Chap. 9. We note that the Markov-additive property would have been preserved in Theorem 6.10 with $(\mathbf{X}, J)$ being a strong MAP and A being an additive functional of J, not necessarily of the form (6.69).

Corollary 6.5. *Suppose $(\mathbf{X}, J_1)$ is an MAP of arrivals on $\mathbf{N}_+^r \times E_1$ with $(Q^1, \{\Lambda_\mathbf{n}^\mathbf{1}\}_{\mathbf{n}>\mathbf{0}})$-source, J_2 is a non-explosive Markov chain on E_2 with generator matrix Q^2, and $(\mathbf{X}, J_1)$ and J_2 are independent. We let $J = (J_1, J_2)$ and assume that for a nonnegative function f on E_2*

$$A(t) = \int_0^t f(J_2(s))\, ds.$$

(a). *With $(\mathbf{X}^A(t), J^A(t)) = (\mathbf{X}(A(t)), J(A(t)))$, the process $(\mathbf{X}^A, J^A)$ is an MAP of arrivals on $\mathbf{N}_+^r \times E_1 \times E_2$ with $(AQ, \{A\Lambda_\mathbf{n}\}_{\mathbf{n}>\mathbf{0}})$-source, where*

$$Q = Q^1 \oplus Q^2, \qquad A = \left(f(j_2)\delta_{(j_1,j_2)(k_1,k_2)}\right), \qquad \Lambda_\mathbf{n} = \Lambda_\mathbf{n}^\mathbf{1} \oplus 0.$$

(b). *Suppose $\inf_{j_2 \in E_2} f(j_2) > 0$, and let $(\mathbf{X}^B(t), J^B(t)) = (\mathbf{X}(B(t)), J(B(t)))$, with $B(t)$ as in (6.70). Then the process $(\mathbf{X}^B, J^B)$ is an MAP of arrivals on $\mathbf{N}_+^r \times E_1 \times E_2$ with source*

$$(A^{-1}Q, \{A^{-1}\Lambda_\mathbf{n}\}_{\mathbf{n}>\mathbf{0}}).$$

Proof. If we define $O(t) = \mathbf{0} \in \mathbf{N}_+^r$, it is easy to see that (O, J_2) is an MAP of arrivals on $\mathbf{N}_+^r \times E_2$ with $(Q^2, \{0\}_{\mathbf{n}>\mathbf{0}})$-source, and is independent of $(\mathbf{X}, J_1)$. Using Theorem 6.9 (b), it then follows that $(\mathbf{X}, J)$ is an MAP of arrivals on $\mathbf{N}_+^r \times E$ with source

$$(Q, \{\Lambda_\mathbf{n}\}_{\mathbf{n}>\mathbf{0}}) = (Q^1 \oplus Q^2, \{\Lambda_\mathbf{n}^\mathbf{1} \oplus 0\}_{\mathbf{n}>\mathbf{0}}).$$

The statements now follow by using Theorem 6.10. □

Neuts [73] considered a random time transformation of particular MRPs associated with the simple BMAP, which is related with the transformations considered in Corollary 6.5; his approach is not applicable to multivariate arrival processes.

If in Corollary 6.5 $\mathbf{X}(= X)$ is a Poisson process with rate λ, $J_2 = J$ is a Markov chain on $\{1, 2, \ldots, m\}$ with generator matrix Q, and X and

J are independent, then (X^A, J^A) is an MMPP with $(AQ, \lambda A)$-source and (X^B, J^B) is an MMPP with $(A^{-1}Q, \lambda^{-1}A)$-source. Thus the transformations described in Theorem 6.10 transform Poisson processes into MMPPs. It is also easy to see that MMPPs are also transformed into MMPPs, and the same is true for the transformations in Theorem 6.9 and Corollary 6.3; this shows that MMPPs have many closure properties.

6.7 Markov-Bernoulli Recording of MAPs of Arrivals

In this section we define a type of secondary recording which includes important special cases of marking and thinning. We consider in particular Markov-Bernoulli recording of MAPs of arrivals, for which the secondary process turns out to be an MAP of arrivals.

Secondary recording of an arrival process is a mechanism that from an *original* arrival process generates a *secondary* arrival process. A classical example of secondary recording is the Bernoulli thinning which records each arrival in the original process, independently of all others, with a given probability p or removes it with probability $1-p$. For MAPs of arrivals $(\mathbf{X}, J)$ we consider secondary recordings that leave J unaffected.

A simple example for which the probability of an arrival being recorded varies with time is the case where there is a recording station (or control process) which is *on* and *off* from time to time, so that arrivals in the original process are recorded in the secondary process during periods in which the station is operational (*on*). The control process may be internal or external to the original process and may be of variable complexity (e.g. rules for access of customer arrivals into a queueing network may be simple or complicated and may depend on the state of the network at arrival epochs or not).

Another example is the one in which the original process counts arrivals of batches of customers into a queueing system, while the secondary process counts the number of individual customers, which is more important than the original arrival process in case service is offered to customers one at a time (this is a way in which the compound Poisson process may be obtained from the simple Poisson process). Here the secondary process usually counts more arrivals than the original process, whereas in the previous examples the secondary process always records fewer arrivals than the original process (which corresponds to thinning).

Secondary recording is related with what is called *marking* of the original process (see e.g. [56]). Suppose that each arrival in the original process is given a *mark* from a space of marks M, independently of the marks given to other arrivals (e.g. for Bernoulli thinning with probability p each point is marked "recorded" with probability p or "removed" with probability $1 - p$). If we consider the process that accounts only for arrivals which have marks on a subset C of M, it may be viewed as a secondary recording of the original process. A natural mark associated with cells moving in a network is the pair of origin and destination nodes of the cell. If the original process accounts for traffic generated in the network, then the secondary process may represent the traffic generated between a given pair of nodes, or between two sets of nodes. The marks may represent priorities. These are commonly used in modeling access regulators to communications network systems. A simple example of an access regulator is the one in which cells (arrivals) which are judged in violation of the "contracts" between the network and the user are marked and may be dropped from service under specific situations of congestion in the network; other cells are not marked and carried through the network (see e.g. [33]). The secondary processes of arrivals of marked and unmarked cells are of obvious interest.

Consider a switch which receives inputs from a number of sources and suppose that the original process accounts for inputs from those sources. If we are interested in studying one of the sources in particular, we should observe a secondary process which accounts only for input from this source; this corresponds to a marginal of the original process. In case all sources generate the same type of input, what is relevant for the study of performance of the system is the total input arriving to the switch; this is a secondary process which corresponds to the sum of the coordinates of the original arrival process. This shows that some of the transformations of arrival processes considered in Sec. 6.6 may also be viewed as special cases of secondary recording of an arrival process.

Suppose that $(\mathbf{X}, J)$ is an MAP of arrivals on $\mathbf{N}_+^r \times E$ with $(Q, \{\Lambda_\mathbf{n}\}_{\mathbf{n}>\mathbf{0}})$-source. We consider a recording mechanism that independently records with probability $r_{jk}(\mathbf{n}, \mathbf{m})$ an arrival in $\mathbf{X}$ of a batch of size $\mathbf{n}$ associated with a transition from j to k in J as an arrival of size $\mathbf{m}$ in the secondary process (with $\mathbf{m} \in \mathbf{N}_+^s$ for some $s \geq 1$). The operation is identified by the set R of recording probabilities

$$R = \{R_{(\mathbf{n},\mathbf{m})} = (r_{jk}(\mathbf{n}, \mathbf{m})) : \mathbf{n} \in \mathbf{N}_+^r, \mathbf{m} \in \mathbf{N}_+^s\} \tag{6.71}$$

where $r_{jk}(\mathbf{n}, \cdot)$ is a probability function on $\mathbf{N}_+^s$, and $r_{jk}(\mathbf{0}, \mathbf{m}) = \delta_{\mathbf{m0}}$. We call this recording *Markov-Bernoulli recording with probabilities in R*

and denote the resulting secondary process as $(\mathbf{X}^R, J)$. Thus $(\mathbf{X}^R, J)$ is a process on $\mathbf{N}_+^s \times E$ which is non-decreasing in $\mathbf{X}^R$, and which increases only when $\mathbf{X}$ increases, i.e.

$$\mathbf{X}^R(t) = \mathbf{X}^R(T_p^\circ) \qquad (T_p^\circ \le t < T_{p+1}^\circ). \tag{6.72}$$

Moreover, for $\mathbf{n} > \mathbf{0}$

$$r_{jk}(\mathbf{n}, \mathbf{m}) = P\{\mathbf{X}^R(T_{p+1}^\circ) - \mathbf{X}^R(T_p^\circ) = \mathbf{m} \mid A_{jk}(\mathbf{n})\} \tag{6.73}$$

with $A_{jk}(\mathbf{n}) = \{\mathbf{X}(T_{p+1}^\circ) - \mathbf{X}(T_p^\circ) = \mathbf{n}, J(T_{p+1}^\circ -) = j, J(T_{p+1}^\circ) = k\}$.

Theorem 6.11. (Markov-Bernoulli recording).

Suppose that $(\mathbf{X}, J)$ is an MAP of arrivals on $\mathbf{N}_+^r \times E$ with $(Q, \{\Lambda_n\}_{n>0})$-source and $R = \{R_{(\mathbf{n},\mathbf{m})}, \mathbf{n} \in \mathbf{N}_+^r, \mathbf{m} \in \mathbf{N}_+^s\}$ is a set of recording probabilities. Then:

(a). *The process $(\mathbf{X}, \mathbf{X}^R, J)$ is an MAP of arrivals on $\mathbf{N}_+^{r+s} \times E$ with source*

$$(Q, \{\Lambda_\mathbf{n} \bullet R_{(\mathbf{n},\mathbf{m})}\}_{(\mathbf{n},\mathbf{m})>\mathbf{0}}). \tag{6.74}$$

(b). *The process $(\mathbf{X}^R, J)$ is an MAP of arrivals on $\mathbf{N}_+^s \times E$ with source*

$$(Q, \{\Lambda_\mathbf{m}^R\}_{\mathbf{m}>\mathbf{0}}) = \left(Q, \left\{\sum_{\mathbf{n}>\mathbf{0}} \Lambda_\mathbf{n} \bullet R_{(\mathbf{n},\mathbf{m})}\right\}_{\mathbf{m}>\mathbf{0}}\right). \tag{6.75}$$

Proof. (a). From (6.72) and (6.73), it is easy to see that $(\mathbf{X}, \mathbf{X}^R, J)$ is an MAP of arrivals on $\mathbf{N}_+^{r+s} \times E$. If $P^\mathbf{X}$ is the probability transition measure of $(\mathbf{X}, J)$, the transition probability measure P of $(\mathbf{X}, \mathbf{X}^R, J)$ is such that for $j, k \in E$ and $(\mathbf{n}, \mathbf{m}) > \mathbf{0}$

$$P_{jk}((\mathbf{n}, \mathbf{m}); h) = P_{jk}^\mathbf{X}(\mathbf{n}; h) r_{jk}(\mathbf{n}, \mathbf{m}) + o(h) \tag{6.76}$$

since the probability of two or more transitions in $(\mathbf{X}, J)$ in time h is of order $o(h)$. In any case, since $r_{jk}(\mathbf{0}, \mathbf{m}) = 0$ for $\mathbf{m} > \mathbf{0}$, (6.76) gives for $(\mathbf{n}, \mathbf{m}) > \mathbf{0}$

$$P_{jk}((\mathbf{n}, \mathbf{m}); h) = \lambda_{jk}(\mathbf{n}) r_{jk}(\mathbf{n}, \mathbf{m}) h + o(h).$$

This implies that in fact $(\mathbf{X}, \mathbf{X}^R, J)$ has $(Q, \{\Lambda_n \bullet R_{(\mathbf{n},\mathbf{m})}\}_{(\mathbf{n},\mathbf{m})>\mathbf{0}})$-source.

(b). The statement follows from (a) using Corollary 6.3 (a). □

We give now two examples of Markov-Bernoulli recording. In Example 6.5 we view the overflow process from a state dependent $M/M/1/K$ system as a special case of secondary recording of the arrival process of customers to the system, and in Example 6.6 we give a more elaborated example of secondary recording.

Example 6.5. (Overflow from a state dependent $M/M/1/K$ queue). We consider a Markov-modulated $M/M/1/K$ system with batch arrivals with (independent) size distribution $\{p_n\}_{n>0}$. When there is an arrival of a batch with n customers and only $m < n$ positions are available only m customers from the batch enter the system. Assume that the service rate is μ_j and the arrival rate of batches is α_j, whenever the number of customers in the system is j. We let $J(t)$ be the number of customers in the system at time t and $X(t)$ be the number of customer arrivals in $(0, t]$.

The process (X, J) is an MAP of arrivals on $\mathbf{N}_+ \times \{0, 1, \ldots, K\}$ with rates

$$\lambda_{jk}(n) = \begin{cases} \alpha_j p_n & k = \min(j+n, K) \\ \mu_j & k = j-1, n = 0 \\ 0 & \text{otherwise} \end{cases} .$$

Its source is $(Q, \{\Lambda_n\}_{n>0})$, where Q is obtained from the matrices $\{\Lambda_n\}_{n\geq 0}$ through (6.8) and (6.9). If we define $X^R(t)$ as the overflow from the system in $(0, t]$, then it is readily seen that (X^R, J) is a Markov-Bernoulli recording of (X, J) with recording probabilities $r_{jk}(n, m) = \mathbf{1}_{\{m=n-(k-j)\}}$ for $n, m > 0$. Thus (X^R, J) is an MAP of arrivals on $\mathbf{N}_+ \times \{0, 1, \ldots, K\}$ with $(Q, \{\Lambda^R_m\}_{m>0})$-source, where $\lambda^R_{jk}(m) = \alpha_j\, \delta_{kK}\, p_{m+(K-j)}$ for $m > 0$.

Example 6.6. Suppose $\mathbf{X} = (X_1, \ldots, X_r)$, $\mathbf{Y} = (Y_1, \ldots, Y_s)$, and $(\mathbf{X}, \mathbf{Y}, J_2)$ is an MAP of arrivals on $\mathbf{N}_+^{r+s} \times E_2$ with $(Q, \{\Lambda_{(\mathbf{n},\mathbf{m})}\}_{(\mathbf{n},\mathbf{m})>\mathbf{0}})$-source. We are interested in keeping only the arrivals in $\mathbf{X}$ which are preceded by an arrival in $\mathbf{Y}$ without a simultaneous arrival in $\mathbf{X}$. Denote by $\mathbf{X}^R$ be the arrival counting process which we obtain by this operation. For simplicity, we assume that E_2 is finite and $\Lambda_{(\mathbf{n},\mathbf{m})} = 0$ if $\mathbf{n} > \mathbf{0}$ and $\mathbf{m} > \mathbf{0}$.

We let $\{T_p^{\mathbf{X}}\}_{p\geq 0}$ and $\{T_p^{\mathbf{Y}}\}_{p\geq 0}$ be the successive arrival epochs in $\mathbf{X}$ and $\mathbf{Y}$, respectively, denote $J = (J_1, J_2)$ with

$$J_1(t) = \begin{cases} 1 & \max\{T_p^{\mathbf{Y}} : T_p^{\mathbf{Y}} \leq t\} > \max\{T_p^{\mathbf{X}} : T_p^{\mathbf{X}} \leq t\} \\ 0 & \text{otherwise} \end{cases} .$$

It can be checked in a routine fashion that $((\mathbf{X}, \mathbf{Y}), J)$ is an MAP of arrivals on $\mathbf{N}_+^{r+s} \times \{0, 1\} \times E_2$ with $(Q^\star, \{\Lambda^\star_{(\mathbf{n},\mathbf{m})}\}_{(\mathbf{n},\mathbf{m})>\mathbf{0}})$-source (with the states of (J_1, J_2) ordered in lexicographic order), where $\Lambda^\star_{(\mathbf{n},\mathbf{m})} = 0$ if $\mathbf{n} > \mathbf{0}$ and $\mathbf{m} > \mathbf{0}$, and

$$Q^\star = \begin{bmatrix} Q - \sum_{\mathbf{m}>\mathbf{0}} \Lambda_{(\mathbf{0},\mathbf{m})} & \sum_{\mathbf{m}>\mathbf{0}} \Lambda_{(\mathbf{0},\mathbf{m})} \\ \sum_{\mathbf{n}>\mathbf{0}} \Lambda_{(\mathbf{n},\mathbf{0})} & Q - \sum_{\mathbf{n}>\mathbf{0}} \Lambda_{(\mathbf{n},\mathbf{0})} \end{bmatrix}, \quad \Lambda^\star_{(\mathbf{n},\mathbf{m})} = \begin{bmatrix} \Lambda_{(\mathbf{n},\mathbf{0})} & \Lambda_{(\mathbf{0},\mathbf{m})} \\ \Lambda_{(\mathbf{n},\mathbf{0})} & \Lambda_{(\mathbf{0},\mathbf{m})} \end{bmatrix}$$

if either $\mathbf{n} = \mathbf{0}$ or $\mathbf{m} = \mathbf{0}$. It is easy to see that $(\mathbf{X}^R, J)$ is a Markov-Bernoulli recording of $((\mathbf{X}, \mathbf{Y}), J)$ with recording probabilities $r_{(p,j)(q,k)}((\mathbf{n}, \mathbf{m}), \mathbf{l}) = \mathbf{1}_{\{p=1, \mathbf{n}=\mathbf{l}\}}$, for $(\mathbf{n}, \mathbf{m}) > \mathbf{0}$. Thus $(\mathbf{X}^R, J)$ is an MAP of arrivals on $\mathbf{N}_+^r \times \{0,1\} \times E_2$ with $(Q^\star, \{\Lambda_\mathbf{n}^R\}_{\mathbf{n}>\mathbf{0}})$-source, where

$$\Lambda_\mathbf{n}^R = \begin{bmatrix} 0 & 0 \\ \Lambda_{(\mathbf{n},\mathbf{0})} & 0 \end{bmatrix}.$$

A particular case of the previous example with $r = 1$ and Y being a Poisson process was considered briefly by Neuts [73]. In addition, He and Neuts [43] considered an instance of the same example in which $(\mathbf{X}, \mathbf{Y}, J_2) = (\mathbf{X}, Y, (K_1, K_2)$ results from patching together independent MAPs $(\mathbf{X}, K_1)$, a general MAP of arrivals, and (Y, K_2), a simple univariate MAP of arrivals. This leads to the recording of arrivals in $\mathbf{X}$ that are immediately preceded by an arrival on the simple univariate MAP of arrivals X.

We now consider the special case of *Markov-Bernoulli marking* for which the space of marks M is discrete. Specifically, each arrival of a batch of size $\mathbf{n}$ associated with a transition from state j to state k in J is given a mark $m \in M$ with *(marking)* probability $c_{jk}(\mathbf{n}, m)$, independently of the marks given to other arrivals. If we assume w.l.o.g. that $M = \{0, 1, \ldots, K\} \subseteq \mathbf{N}_+$, Markov-Bernoulli marking becomes a special case of Markov-Bernoulli recording for which the *marked process* $(\mathbf{X}, \mathbf{X}^R, J)$ is such that $(\mathbf{X}^R, J)$ is a Markov-Bernoulli recording of $(\mathbf{X}, J)$ with recording probabilities

$$R_{(\mathbf{n},m)} = (r_{jk}(\mathbf{n}, m)) = (\mathbf{1}_{\{m \in M\}} c_{jk}(\mathbf{n}, m))$$

for $\mathbf{n} > \mathbf{0}$ and $m \in \mathbf{N}_+$.

Theorem 6.12. *With the conditions described we have the following.*
(a). (**Markov-Bernoulli marking**).

The marked process $(\mathbf{X}, \mathbf{X}^R, J)$ is an MAP of arrivals on $\mathbf{N}_+^{r+1} \times E$ with source

$$(Q, \{\Lambda_\mathbf{n} \bullet R_{(\mathbf{n},m)}\}_{(\mathbf{n},m)>\mathbf{0}}). \tag{6.77}$$

(b). (**Markov-Bernoulli colouring**).

Suppose that the set of marks is finite, i.e. $K < \infty$. We identify mark $m \in M$ as colour m, and let $\mathbf{X}^{(m)}$ be the counting process of arrivals of batches coloured m, for $0 \le m \le K$. The process $(\mathbf{X}^{(0)}, \ldots, \mathbf{X}^{(K)}, J)$ is an MAP of arrivals on $\mathbf{N}_+^{(K+1)r} \times E$ with $(Q, \{\Lambda_\mathbf{a}^\star\}_{\mathbf{a}>\mathbf{0}})$-source. Here for $\mathbf{a} = (\mathbf{a}^{(0)}, \ldots, \mathbf{a}^{(K)}) > \mathbf{0}$ with $\mathbf{a}^{(m)} \in \mathbf{N}_+^r (0 \le m \le K)$

$$\Lambda_\mathbf{a}^\star = \left(\sum_{m=0}^{K} \mathbf{1}_{\{A_\mathbf{a} = \{\mathbf{a}^{(m)}\}\}} \Lambda_{\mathbf{a}^{(m)}} \bullet R_{(\mathbf{a}^{(m)}, m)} \right) \tag{6.78}$$

where $A_{\mathbf{a}} = \{\mathbf{a}^{(l)} : 0 \leq l \leq K, \mathbf{a}^{(l)} > \mathbf{0}\}$.

Proof. The statement (a) is a consequence of Theorem 6.11 (a). The statement (b) follows from (a), Theorem 6.11 (b), and the fact that $(\mathbf{X}^{(0)}, \ldots, \mathbf{X}^{(K)}, J)$ is a Markov-Bernoulli recording of $(\mathbf{X}, \mathbf{X}^R, J)$ with recording probability matrices

$$R^{\star}_{((\mathbf{n},m)(\mathbf{n}^{(0)},\ldots,\mathbf{n}^{(K)}))} = \left(\mathbf{1}_{\{\mathbf{n}^{(m)}=\mathbf{n},\mathbf{n}^{(l)}=\mathbf{0}(0\leq l\leq K, l\neq m)\}}\right)$$

for $0 \leq m \leq K$ and $\mathbf{n}, \mathbf{n}^{(0)}, \ldots, \mathbf{n}^{(K)} \in \mathbf{N}_+^r$, with $(\mathbf{n}, m) > \mathbf{0}$. □

Theorem 6.12 (b) is an extension of the Colouring Theorem for Poisson processes; see ([56], Section 5.1]). An important consequence of Theorem 6.12 is the following result.

Corollary 6.6. *Suppose* $(\mathbf{X}, J)$ *is an MAP of arrivals on* $\mathbf{N}_+^r \times E$ *with* $(Q, \{\Lambda_{\mathbf{n}}\}_{\mathbf{n}>\mathbf{0}})$*-source, and* $\{B_m\}_{0\leq m\leq K}$ *is a finite partition of* $\mathbf{N}_+^r - \{\mathbf{0}\}$. *If for* $0 \leq m \leq K$ *we let* $\mathbf{X}^{(m)}$ *be the counting process of arrivals with batch sizes in* B_m, *then* $(\mathbf{X}^{(0)}, \ldots, \mathbf{X}^{(K)}, J)$ *is an MAP of arrivals on* $\mathbf{N}_+^{(K+1)r} \times E$ *with* $(Q, \{\Lambda^{\star}_{\mathbf{a}}\}_{\mathbf{a}>\mathbf{0}})$*-source, where for* $\mathbf{a} = (\mathbf{a}^{(0)}, \ldots, \mathbf{a}^{(K)}) > \mathbf{0}$, *with* $\mathbf{a}^{(m)} \in \mathbf{N}_+^r$ $(0 \leq m \leq K)$

$$\Lambda^{\star}_{\mathbf{a}} = \begin{cases} \Lambda_{\mathbf{a}^{(m)}} & A_{\mathbf{a}} = \{\mathbf{a}^{(m)}\} \subseteq B_m \\ 0 & \text{otherwise} \end{cases} \tag{6.79}$$

where $A_{\mathbf{a}} = \{\mathbf{a}^{(l)} : 0 \leq l \leq K, \mathbf{a}^{(l)} > \mathbf{0}\}$.

Proof. The statement follows from Theorem 6.12 (b) by considering a Markov-Bernoulli marking of $(\mathbf{X}, J)$ with marking probabilities $c_{jk}(\mathbf{n}, m) = \mathbf{1}_{\{\mathbf{n}\in B_m\}}$. □

Note that the processes $(\mathbf{X}^{(p)}, J)$ and $(\mathbf{X}^{(q)}, J)$ with $p \neq q$ have no common arrival epochs a.s. When J has more than one state, $\mathbf{X}^{(p)}$ and $\mathbf{X}^{(q)}$ $(0 \leq p, q \leq K)$ are not independent, except perhaps in very special cases. In case J has only one state and $r = 1$, X is a compound Poisson process and Corollary 6.6 implies that $X^{(m)} (0 \leq m \leq K)$ are independent compound Poisson processes.

When the primary and secondary processes are defined on the same state space and the recording probabilities are such that $R_{(\mathbf{n},\mathbf{m})} = 0$ for $\mathbf{m} > \mathbf{n}$ we say that the associated secondary recording is a *thinning*, and the secondary process is a *thinned process.* If $\mathbf{X}$ and $\mathbf{X}^R$ are the counting processes of arrivals in the original and the thinned process we may also consider the process $\mathbf{X}^L = \mathbf{X} - \mathbf{X}^R$ which counts lost arrivals. From Theorem 6.11 we know that the *loss* process $(\mathbf{X}^L, J)$ is also an MAP of arrivals, since we may interchange recorded arrivals with non-recorded arrivals. However for Markov-Bernoulli thinning we are able to give sharper results.

Theorem 6.13. (Markov-Bernoulli thinning).

Suppose that $(\mathbf{X}, J)$ *is an MAP of arrivals on* $\mathbf{N}_+^r \times E$ *with* $(Q, \{\Lambda_{\mathbf{n}}\}_{\mathbf{n}>0})$*-source and* $R = \{R_{(\mathbf{n},\mathbf{m})}\}$ *is a set of thinning probabilities. Then the process* $(\mathbf{X}^R, \mathbf{X}^L, J)$ *is an MAP of arrivals on* $\mathbf{N}_+^{2r} \times E$ *with source*

$$(Q, \{\Lambda_{\mathbf{n}+\mathbf{m}} \bullet R_{(\mathbf{n}+\mathbf{m},\mathbf{n})}\}_{(\mathbf{n},\mathbf{m})>\mathbf{0}}). \tag{6.80}$$

Proof. Using arguments similar to the ones in the proof of Theorem 6.11 (a), we may conclude that the process $(\mathbf{X}, \mathbf{X}^R, \mathbf{X}^L, J)$ is an MAP of arrivals on $\mathbf{N}_+^{3r} \times E$ with source

$$\left(Q, \{[\Lambda_{\mathbf{n}} \bullet R_{(\mathbf{n},\mathbf{m})}]\mathbf{1}_{\{\mathbf{p}=\mathbf{n}-\mathbf{m}\}}\}_{(\mathbf{n},\mathbf{m},\mathbf{p})>\mathbf{0}}\right). \tag{6.81}$$

Using Corollary 6.3 (a), the statement follows. □

Example 6.7. Suppose that X is a Poisson process with rate λ, J is a stable finite Markov chain with generator matrix Q, and X and J are independent. We view J as the state of a recording station, and assume that each arrival in X is recorded, independently of all other arrivals, with probability p_j or non-recorded with probability $1-p_j$, whenever the station is in state j. We let $X^R(t)$ be the number of recorded arrivals in $(0, t]$.

Using Theorem 6.9 (b), Theorem 6.13, and Corollary 6.3 (a), we may conclude that (X^R, J) is an MMPP with $(Q, (\lambda p_j \delta_{jk}))$-source. In the special case where the probabilities p_j are either 1 or 0, the station is *on* and *off* from time to time, with the distributions of the on and off periods being Markov dependent.

Chapter 7

Markov-Modulated Single Server Queueing Systems

In this chapter we consider single server queueing systems that are modulated by a discrete time Markov chain on a countable state space. The underlying stochastic process is a Markov random walk (MRW) whose increments can be expressed as differences between service times and interarrival times. We derive the joint distributions of the waiting and idle times in the presence of the modulating Markov chain. Our approach is based on properties of the ladder sets associated with this MRW and its time-reversed counterpart. The special case of a Markov-modulated M/M/1 queueing system is then analyzed and results analogous to the classical case are obtained.

7.1 Introduction

In many practical situations, the customer arrivals and the service mechanism of a queueing system are both influenced by some factors external or internal to the system. This has motivated considerable interest in the study of Markov-modulated queueing models where the influence of these factors is conveyed through a secondary Markov chain.

In most of the models proposed so far the secondary Markov chain is formulated in continuous time. Prabhu and Zhu [96] have given a comprehensive survey of the literature in this area and also investigated a Markov-modulated M/G/1 system. Asmussen [10] has also considered this M/G/1 system using a different approach. In another class of models the secondary

[7]This chapter is an updated version of material included in the article by N.U. Prabhu and L.C. Tang, 'Markov-modulated single-server queueing systems', *J. Appl. Probab.* **31A**, pp. 169–184, 1994, and contents are reproduced with permission of the publisher, the Applied Probability Trust.

Markov chain is formulated in discrete time. Specifically, let us denote by $\{u_k, k \geq 1\}$ and $\{v_k, k \geq 1\}$ the interarrival and service times respectively, $U_n = \sum_{k=1}^{n} u_k$ and $V_n = \sum_{k=1}^{n} v_k$ $(n \geq 1)$. The basic assumption is that $\{(U_n, V_n, J_n)\}$ is a Markov renewal process (MRP), the secondary Markov chain being $J = \{J_n\}$. This implies that the distributions of both the interarrival and the service times of a customer are determined by the J-chain at the epoch of the customer's arrival, in contrast to the case of a continuous-time J where the presence of the secondary Markov chain is felt even between arrival and departure epochs. The queue discipline is first come, first served. The first paper to treat a model of this latter type was by Arjas [5], who obtained some results using a matrix version of the Miller-Kemperman identity. Further work can be found in Takács [109, 110] where some recurrence equations in Banach algebras were derived.

In this chapter, we study this model by considering the underlying MRW $\{(S_n, J_n)\}$ where $S_0 = 0$ and $S_n = V_n - U_n$ $(n \geq 1)$. Here the interarrival and the service times need not be conditionally independent of each other, given the state of J, as we are only considering their difference. The MRW considered here is assumed to be nondegenerate (see Sec. 4.3 for definition and discussion of degeneracy). Essentially, we are extending the techniques used in the classical case (where the state space of J consists of a single element) to the Markov-modulated case. For the purpose of this extension we need several new ideas, including the following:

(a) The class of nondegenerate MRW we consider here is one whose increments X_k can be represented as the differences of two positive random variables. We denote this class as $\mathcal{C}_Q$.
(b) Associated with the given system Σ is what we have designated as the adjoint system $\widehat{\Sigma}$ obtained by a time-reversal of the MRW. This latter MRW also belongs to the class $\mathcal{C}_Q$ and is an extension of the dual for the classical random walks mentioned in Feller ([36], Section XII.2). This system $\widehat{\Sigma}$ is not the the dual of the system considered by Prabhu [86] for the classical case. In the dual considered by Arjas [5], in addition to time-reversal, the sign of the increments of the given MRW are also reversed; thus this dual is different from our adjoint. The result of Lemma 1 indicates that it is our notion of $\widehat{\Sigma}$ that gives the result analogous to that of the classical case; see Prabhu ([86], Lemma 1, p. 24).
(c) Our starting point is Presman's [97] Wiener-Hopf factorization which we have expressed (see Sec. 4.4) in terms of the ladder sets associated

with the system Σ and $\widehat{\Sigma}$. In order to apply this factorization we introduce in Sec. 7.2 notations and results, specifically concerning the distributions of the maximum and minimum functionals of the MRW.

(d) For application to the special case of the Markov-modulated M/M/1 system we have extended the partial lack of memory property used by Prabhu [86] in the classical case. This property is expressed in terms of the increments of the given MRW and its time-reversed counterpart.

In Sec. 7.3 we show that the distributions of the waiting time and the accumulated idle time jointly with the J-chain can be expressed in terms of the ladder measures of the MRWs corresponding to the given system Σ and its adjoint $\widehat{\Sigma}$. We then obtain effortlessly the joint transform of the steady state waiting time, accumulated idle time, and the J-chain.

The special case of a Markov-modulated M/M/1 queue is investigated in Sec. 7.4. Using the partial lack of memory property mentioned in (d) above, we arrive at results analogous to those in the classical M/M/1 queue concerning the idle period. We then derive a matrix identity for this queueing system.

7.2 Preliminary Results

Consider a single server queueing system whose customers belong to certain types indexed by $j \in E$, the switching mechanism between different types of customers being governed by a Markov chain J with state space E. Each customer type has his own interarrival time and service time distributions determined by the state of the J chain at the epoch of arrival. The Markov chain J is assumed to be irreducible and persistent. The queue discipline is first come, first served.

We denote by $\{u_k, k \geq 1\}$ the sequence of interarrival times and by $\{v_k, k \geq 1\}$ the sequence of service times. Let $X_k = v_k - u_k$ $(k \geq 1)$, $S_0 = 0$ and $S_n = \sum_{k=1}^{n} X_k$ $(n \geq 1)$. Then $(S, J) = \{S_n, J_n; n \geq 0\}$ is a time homogeneous MRW. It belongs to the class $\mathcal{C}_Q$ defined as follows.

Definition 7.1. The class $\mathcal{C}_Q$ of MRWs is the one in which the increments X_k of the MRW can be represented by the difference of two positive random variables.

The transition distribution measure of this MRW is given by

$$Q_{jk}^{(n)}\{A-x\} = P\{(S_{m+n}, J_{m+n}) \in A \times \{k\} | (S_m, J_m) = (x, j)\}$$
$$= P\{(S_{m+n} - S_m, J_{m+n}) \in (A-x) \times \{k\} | J_m = j\} \quad (7.1)$$

for $j, k \in E$, $x \in \mathbf{R}$, and a Borel subset A of $\mathbf{R}$. Clearly, the state space of the MRW is $\mathbf{R} \times E$. We assume that the MRW is nondegenerate. As usual, we write Q_{jk} for $Q_{jk}^{(1)}$ and denote the conditional probabilities and expectations given $J_0 = j$ as P_j and E_j respectively.

It is clear that the MRW resulting from a given Markov-modulated queue is unique. Conversely from the uniqueness (up to translation) of the difference between two positive random variables, given a MRW in class $\mathcal{C}_Q$, there exists an Markov-modulated queueing system, say Σ, which is also unique (up to translation).

In addition, consider the time-reversed version $\{(\widehat{S}_n, \widehat{J}_n), n \geq 0\}$ of the given MRW whose transition probability measure is given by

$$\widehat{Q}_{jk}\{A\} = \frac{\pi_k}{\pi_j} Q_{kj}\{A\} \qquad (j, k \in E) \quad (7.2)$$

where $Q_{kj}\{\cdot\}$ is the transition probability measure of (S, J), $\{\pi_i\}$ is a sequence of positive numbers, and A is a Borel subset of $\mathbf{R}$.

Consider the queueing system $\widehat{\Sigma}$ whose associated MRW has the transition measure given by (7.2). This system is unique up to translation, in view of the fact that $X_k = (v_k + c) - (u_k + c)$ for any $c \geq 0$. We call $\widehat{\Sigma}$ the *adjoint system* of the given system Σ. We shall use $\widehat{(\cdot)}$ to denote the counterpart of $(\cdot)$ for the adjoint system. By setting $A = \mathbf{R}$ in the above equation, we see that

$$\pi_j \widehat{P}_{jk} = \pi_k P_{kj},$$

which shows that $\{\pi_j\}$ is necessarily the stationary measure of both J and $\widehat{J}$.

We denote W_n as the waiting time of the nth customer, I_n as the idle time (if any) that just terminates upon the arrival of this customer, and $\mathcal{I}_n = I_1 + I_2 + \cdots + I_n$ the accumulated idle time. It is clear that W_n and $\mathcal{I}_n$ are related to S_n and m_n in the same way as in the classical case; thus for $W_0 = 0$ we have

$$W_n = S_n - m_n \quad \text{and} \quad \mathcal{I}_n = -m_n \quad (7.3)$$

where $m_n = \min_{0 \leq k \leq n} S_k$ $(n \geq 0)$ is the minimum functional of the underlying MRW. Similarly, we denote the maximum functional by $M_n = \max_{0 \leq k \leq n} S_k$.

In order to obtain the distribution of W_n and $\mathcal{I}_n$, as represented by (7.3), we shall first recall that Lemma 4.4 establishes the following relation between the extremum functionals of the given MRW and the MRW of $\widehat{\Sigma}$:

$$\begin{aligned} P_k\{S_n - m_n \le x, -m_n \le y, J_n = j\} \\ = \frac{\pi_j}{\pi_k} P_j\{\widehat{M}_n \le x, \widehat{M}_n - \widehat{S}_n \le y, \widehat{J}_n = k\} \end{aligned} \tag{7.4}$$

for $n \ge 1$, $x, y \ge 0$, and $j, k \in E$.

We now define the following semirecurrent sets associated with the extremum functionals

$$\begin{cases} \zeta_+ = \{(n,j) : S_n = M_n, J_n = j, \rho_n = n\} \\ \zeta_- = \{(n,j) : S_n = m_n, J_n = j\} \end{cases} \tag{7.5}$$

where ρ_n is the first epoch at which the MRW attains the value M_n.

Associated with these sets are the following measures. Let

$$\begin{cases} u_{jk}^{(0)}(x) = \delta_{jk} \mathbf{1}_{\{x \ge 0\}} \\ u_{jk}^{(n)}(x) = P_j\{(n,k) \in \zeta_+, S_n \le x\} \quad (n \ge 1) \end{cases} \tag{7.6}$$

$$\begin{cases} v_{jk}^{(0)}(x) = \delta_{jk} \mathbf{1}_{\{x \le 0\}} \\ v_{jk}^{(n)}(x) = P_j\{(n,j) \in \zeta_-, S_n \ge x\} \quad (n \ge 1) \end{cases} \tag{7.7}$$

$$u_{jk}(x) = \sum_{n=0}^{\infty} u_{jk}^{(n)}(x) \qquad \text{and} \qquad v_{jk}(x) = \sum_{n=0}^{\infty} v_{jk}^{(n)}(x). \tag{7.8}$$

Note that $u_{jk}^{(n)}(\cdot)$ and $u_{jk}(\cdot)$ are concentrated on $\mathbf{R}_+ = [0, \infty)$, whereas $v_{jk}^{(n)}(\cdot)$ and $v_{jk}(\cdot)$ are concentrated on $\mathbf{R}_- = (-\infty, 0]$. These measure are all finite. To see this we define the two sequences of random variables

$$\begin{cases} T_r = \min\{n > T_{r-1} : S_n > S_{T_{r-1}}\} \\ \bar{T}_r = \min\{n > \bar{T}_{r-1} : S_n \le S_{\bar{T}_{r-1}}\} \end{cases} \quad (r \ge 1) \tag{7.9}$$

with $T_0 = \bar{T}_0 = 0$. In the queueing system Σ, $\bar{T}_k - \bar{T}_{k-1}$ represents the number of customers served during the kth busy-period. It is clear that the collection of these ladder heights with the J state are exactly the two sets ζ_+ and ζ_-; i.e.

$$\begin{cases} \zeta_+ = \{(T_0, J_0), (T_1, J_{T_1}), \dots\} \\ \zeta_- = \{(\bar{T}_0, J_0), (\bar{T}_1, J_{\bar{T}_1}), \dots\} \end{cases} \tag{7.10}$$

It turns out that (7.8) are Markov renewal measures of the MRPs $\{(S_{T_k}, J_{T_k})\}$ and $\{(-S_{\bar{T}_k}, J_{\bar{T}_k})\}$ respectively. It also follows from Markov renewal theory that $\sum_{k\in E} u_{jk}(x) < \infty$ for all finite $x \in \mathbf{R}_+$. For the adjoint system these measures are related to the number of customers served in a busy-period in the following manner, in view of Lemma 4.5. For $x > 0$, $y \leq 0$, $n \in \mathbf{N}_+$, and $j, k \in E$:

$$\widehat{u}_{jk}^{(n)}(x) = \frac{\pi_k}{\pi_j} P_k\{\bar{T}_1 > n, S_n \leq x, J_n = j\} \tag{7.11}$$

$$\widehat{v}_{jk}^{(n)}(y) = \frac{\pi_k}{\pi_j} P_k\{T_1 > n, S_n \geq y, J_n = j\}. \tag{7.12}$$

We can also express the joint distributions of the extremum functionals as functions of these measures. Namely, from Theorem 4.6, we have for $x, y \geq 0$, $n \in \mathbf{N}_+$, and $i, k \in E$:

$$\begin{aligned} P_i\{M_n \leq x, M_n - S_n \leq y, J_n = k\} \\ = \pi_k \sum_{j\in E} \sum_{m=0}^{n} u_{ij}^{(m)}(x) \frac{1}{\pi_j} \widehat{v}_{kj}^{(n-m)}(-y) \end{aligned} \tag{7.13}$$

$$\begin{aligned} P_i\{S_n - m_n \leq x, -m_n \leq y, J_n = k\} \\ = \pi_k \sum_{j\in E} \sum_{m=0}^{n} \widehat{u}_{kj}^{(m)}(x) \frac{1}{\pi_j} v_{ij}^{(n-m)}(-y). \end{aligned} \tag{7.14}$$

7.3 Waiting and Idle Times

In this section we obtain the joint distribution of the waiting time and accumulated idle time and its transform. We also give explicit expressions for their limit distributions.

Theorem 7.1. *For $x, y \geq 0$, $n \in \mathbf{N}_+$, and $i, k \in E$, we have*

$$P_i\{W_n \leq x, \mathcal{I}_n \leq y, J_n = k\} = \pi_k \sum_{j\in E} \sum_{m=0}^{n} \widehat{u}_{kj}^{(m)}(x) \frac{1}{\pi_j} v_{ij}^{(n-m)}(-y). \tag{7.15}$$

Moreover,

$$\lim_{n\to\infty} P_i\{W_n \leq x, J_n = k\} = \pi_k \sum_{j\in E} \widehat{u}_{kj}(x) P_j\{\widehat{T}_1 = \infty\} \tag{7.16}$$

$$\lim_{n\to\infty} P_i\{\mathcal{I}_n \leq y, J_n = k\} = \pi_k \sum_{j\in E} v_{ij}(-y) P_j\{\bar{T}_1 = \infty\}. \tag{7.17}$$

Proof. From (7.3) and (7.14), we find that

$$\begin{aligned} P_i\{W_n \leq x, \mathcal{I}_n \leq y, J_n = k\} &= P_i\{S_n - m_n \leq x, -m_n \leq y, J_n = k\} \\ &= \pi_k \sum_{j\in E} \sum_{m=0}^{n} \widehat{u}_{kj}^{(m)}(x) \frac{1}{\pi_j} v_{ij}^{(n-m)}(-y) \end{aligned}$$

as desired. Letting $y \to \infty$ we obtain

$$\lim_{n\to\infty} P_i\{W_n \leq x, J_n = k\} = \lim_{n\to\infty} \pi_k \sum_{j\in E} \sum_{m=0}^{n} \widehat{u}_{kj}^{(m)}(x) \frac{1}{\pi_j} v_{ij}^{(n-m)}(-\infty).$$

Also, in view of (7.12),

$$\begin{aligned}
\lim_{n\to\infty} \widehat{v}_{ij}^{(n)}(-\infty) &= \lim_{n\to\infty} P_i\{(n,j) \in \widehat{\zeta}_-, \widehat{S}_n \geq -\infty\} \\
&= \lim_{n\to\infty} \frac{\pi_j}{\pi_i} P_j\{T_1 > n, S_n \geq -\infty, J_n = i\} \\
&= \frac{\pi_j}{\pi_i} \cdot \pi_i P_j\{T_1 = \infty\}
\end{aligned}$$

where the last equality follows from the fact that the event $\{T_1 = \infty\} \in \cup_r \sigma\{X_1, X_2, \dots, X_r\}$, and the event $\{J_\infty = i\}$ belongs to the tail σ-field which is independent of the former for each fixed j. As a result, we obtain

$$\begin{aligned}
\lim_{n\to\infty} P_i\{W_n \leq x, J_n = k\} &= \pi_k \sum_{j\in E} \sum_{m=0}^{\infty} \widehat{u}_{kj}^{(m)}(x) \frac{1}{\pi_j} \lim_{r\to\infty} v_{ij}^{(r)}(-\infty) \\
&= \pi_k \sum_{j\in E} \widehat{u}_{kj}(x) P_j\{\widehat{T}_1 = \infty\}
\end{aligned}$$

which proves (7.16). The proof of (7.17) is similar, making $x \to \infty$ and using (7.11). □

It can be seen that the limit distributions of W_n and $\mathcal{I}_n$ can be interpretated as the weighted average of the measures $\widehat{u}_{kj}(x)$ with weights $P_j\{\widehat{T}_1 = \infty\}$ and $v_{ij}(-y)$ with weights $P_j\{\bar{T}_1 = \infty\}$ respectively. Both distributions are independent of J_0. The following result shows that the contributions of these weights are "mutually exclusive."

Proposition 7.1. *We have*

$$P_j\{\bar{T} = \infty\} P_j\{\widehat{T} = \infty\} = 0 \qquad (j \in E) \tag{7.18}$$

i.e. $\bar{T}$ *and* $\widehat{T}$ *cannot be both defective.*

Proof. First we recall that from Theorem 4.15, we have

$$\mathbf{I} - z\Phi(\omega) = \mathbf{D}_\pi^{-1}[\mathbf{I} - \widehat{\chi}^+(z,\omega)]'\, \mathbf{D}_\pi[\mathbf{I} - \chi^-(z,\omega)] \tag{7.19}$$

for $|z| \leq 1$ and $\omega \in \mathbf{R}$, where $\mathbf{D}_\pi = \text{diag}(\pi)$, the prime denotes transpose, $\Phi(\omega) = (\phi_{jk}(\omega))$, $\chi^+(z,\omega) = (\chi^+_{jk}(z,\omega))$, and $\chi^-(z,\omega) = (\chi^-_{jk}(z,\omega))$ are matrices with indices on $E \times E$, given by

$$\phi_{jk}(\omega) = \int_{-\infty}^{\infty} e^{i\omega x} Q_{jk}\{dx\} \tag{7.20}$$

$$\chi^+_{jk}(z,\omega) = E_j(z^T e^{i\omega S_T}; J_T = k) \tag{7.21}$$

$$\chi^-_{jk}(z,\omega) = E_j(z^{\bar{T}} e^{i\omega S_{\bar{T}}}; J_{\bar{T}} = k) \tag{7.22}$$

and as usual, $T = T_1$, $\bar{T} = \bar{T}_1$, while $\widehat{\chi}^+(z,\omega)$ and $\widehat{\chi}^-(z,\omega)$ denote the counterparts for the corresponding time-reversed MRW.

Setting $\omega = 0$ in (7.19) we obtain

$$I - zP = \mathbf{D}_\pi^{-1}[I - (E_j(z^{\widehat{T}}; J_{\widehat{T}} = k))]'\mathbf{D}_\pi[I - (E_j(z^{\bar{T}}; J_{\bar{T}} = k))].$$

Post-multiplying both sides of this last identity by $\mathbf{e} = (1,1,\ldots,1)'$ and letting $z \to 1$, we arrive at

$$\mathbf{D}_\pi(\ldots,0,\ldots)' = [I - (P_j\{\widehat{T} < \infty; J_{\widehat{T}} = k\})]'\mathbf{D}_\pi(\ldots,1 - P_j\{\bar{T} < \infty\},\ldots)'.$$

Premultiplying both sides of this equation by $\mathbf{e}'$ we obtain

$$0 = \pi_j P_j\{\widehat{T} = \infty\} P_j\{\bar{T} = \infty\},\ \forall j. \qquad \square$$

The following classification of the underlying MRW shows that the existence of the limit distributions in (7.16)-(7.17) is mutually exclusive. The classification is a direct consequence of Corollary 4.1 and Theorem 4.11.

Proposition 7.2. *Suppose that the underlying MRW of Σ is nondegenerate with persistent J. Then we have*

(a) $P_j\{\widehat{T} = \infty\} = P_j\{\bar{T} = \infty\} = 0,\ \forall j$

$$\Leftrightarrow \liminf_n S_n = -\infty,\ \limsup_n S_n = +\infty \Leftrightarrow \mu = 0$$

(b) $P_j\{\widehat{T} = \infty\} = 0,\ \forall j,$ *and* $P_j\{\bar{T} = \infty\} > 0,$ *for some* j

$$\Leftrightarrow \lim_n S_n = +\infty \Leftrightarrow \mu > 0$$

(c) $P_j\{\widehat{T} = \infty\} > 0$ *for some* $j,$ *and* $P_j\{\bar{T} = \infty\} = 0,\ \forall j$

$$\Leftrightarrow \lim_n S_n = -\infty \Leftrightarrow \mu < 0$$

where $\mu = \sum_{j \in E} \pi_j E_j(X_1)$*, provided that* $E_j(X_1)$ *exists for all* $j \in E$*, is the aggregated mean.*

Note that $P_j\{T = \infty\} = P_j\{\widehat{T} = \infty\}$ and, similarly, $P_j\{\bar{T} = \infty\} = P_j\{\widehat{\bar{T}} = \infty\}$. In addition, $P_j\{T = \infty\} = 0\ (> 0)$ corresponds to the case where ζ^+ is nonterminating (terminating) and $P_j\{\bar{T} = \infty\} = 0\ (> 0)$ corresponds to the case where ζ^- is nonterminating (terminating).

From (7.15), it is easy to obtain the transform of $\{(W_n, \mathcal{I}_n, J_n)\}$ as the transforms of the measures $\widehat{u}_{jk}^{(n)}(x), v_{jk}^{(n)}(-y)$ are easy to evaluate.

Theorem 7.2. *For* $j,k \in E$, $|z| \le 1$, $\omega_1, \omega_2 \in \mathbf{R}$, *let*

$$\Psi_{jk}(z,\omega_1,\omega_2) = \sum_{n=0}^{\infty} z^n E_j(e^{i\omega_1 W_n + i\omega_2 \mathcal{I}_n}; J_n = k). \qquad (7.23)$$

Then, we have

$$\Psi(z,\omega_1,\omega_2) = [\mathbf{I} - \chi^-(z,-\omega_2)]^{-1}[\mathbf{I} - \chi^-(z,\omega_1)][\mathbf{I} - z\Phi(\omega_1)]^{-1}. \qquad (7.24)$$

Proof. From (7.15) we obtain

$$\sum_{n=0}^{\infty} z^n E_j(e^{i\omega_1 W_n + i\omega_2 \mathcal{I}_n}; J_n = k) = \sum_{n=0}^{\infty} z^n \int_0^{\infty}\int_0^{\infty} \pi_k \sum_{l\in E}\sum_{m=0}^{n} \widehat{u}_{kl}^{(m)}(dx) \cdot \frac{1}{\pi_l} v_{jl}^{(n-m)}(-dy) e^{i\omega_1 x + i\omega_2 y}$$
$$= [v^*(z, -\omega_2)\mathbf{D}_\pi^{-1}\widehat{u}^*(z,\omega_1)'\mathbf{D}_\pi]_{jk}$$

where

$$\widehat{u}_{jk}^*(z,\omega) = \sum_{n=0}^{\infty} z^n \int_0^{\infty} e^{i\omega x}\widehat{u}_{jk}^{(n)}(dx)$$

and

$$v_{jk}^*(z,\omega) = \sum_{n=0}^{\infty} z^n \int_{-\infty}^{0} e^{i\omega x} v_{jk}^{(n)}(dx).$$

Since

$$\widehat{u}_{jk}^*(z,\omega) = \sum_{m=0}^{\infty} E_j[z^{\widehat{T}_m} e^{i\omega S_{\widehat{T}_m}}; J_{\widehat{T}_m} = k] = \sum_{m=0}^{\infty} [(\widehat{\chi}^+(z,\omega))^m]_{jk}$$

we find that

$$\widehat{u}^*(z,\omega) = [\mathbf{I} - \widehat{\chi}^+(z,\omega)]^{-1}$$

and similarly

$$v^*(z,\omega) = [\mathbf{I} - \chi^-(z,\omega)]^{-1}.$$

Moreover, from (7.19) we have

$$\mathbf{D}_\pi^{-1}[\mathbf{I} - \widehat{\chi}^+(z,\omega)']^{-1}\mathbf{D}_\pi = [\mathbf{I} - \chi^-(z,\omega)][\mathbf{I} - z\Phi(\omega)]^{-1}.$$

Combining this with the above results we arrive at (7.24). □

It can be seen that if we set $\omega_1 = -\omega_2$, we get

$$\Psi(z,\omega_1,-\omega_1) = [\mathbf{I} - z\Phi(\omega_1)]^{-1}$$

which is expected since $W_n - \mathcal{I}_n = S_n$.

Similarly, from (7.16) we obtain the following.

Theorem 7.3. *Suppose that $\mu < 0$ so that $(W,J) = \lim_n (W_n, J_n)$ exists. Then $E_j(e^{i\omega W}; J = k)$ is the kth entry of the vector*

$$\mathbf{D}_\pi[\mathbf{I} - \widehat{\chi}^+(1,\omega)]^{-1}[\mathbf{I} - \widehat{\chi}^+(1,0)]\mathbf{e} \tag{7.25}$$

and this is independent of J_0.

Proof. From (7.16) we find that

$$\begin{aligned} E_j(e^{i\omega W}; J = k) &= \pi_k \sum_{l \in E} \int_0^\infty e^{i\omega x} \widehat{u}_{kl}(dx) P_l\{\widehat{T} = \infty\} \\ &= \pi_k \sum_{l \in E} P_l\{\widehat{T} = \infty\} \widehat{u}^*_{kl}(1, \omega). \end{aligned}$$

The desired result follows from the facts that $\widehat{u}^*(z,\omega) = [\mathbf{I} - \widehat{\chi}^+(z,\omega)]^{-1}$ and $P_l\{\widehat{T} = \infty\}$ is the lth entry of $[\mathbf{I} - \widehat{\chi}^+(1,0)]\mathbf{e}$. □

Next we give the limit distributions of the waiting and accumulated idle times when $\mu > 0$ and $\mu < 0$, respectively, in which case we do not have steady state distributions for these random variables. However, as in the classical system, one may use the central limit result for the underlying Markov random walk so that the waiting and accumulated times, when suitably normed, have limit distributions. We let $\mathcal{N}$ denote the standard normal distribution and $\xrightarrow{\mathcal{L}}$ denote convergence in distribution. The constant σ appearing next is the one given in Theorem 5.1.

Theorem 7.4. (i). *If* $\mu > 0, \sigma < \infty$, *we have*

$$\frac{W_n - n\mu}{\sigma\sqrt{n}} \xrightarrow{\mathcal{L}} \mathcal{N}. \tag{7.26}$$

(ii). *If* $\mu < 0, \sigma < \infty$, *we have*

$$\frac{\mathcal{I}_n + n\mu}{\sigma\sqrt{n}} \xrightarrow{\mathcal{L}} \mathcal{N}. \tag{7.27}$$

Proof. From (7.3) we have

$$\frac{W_n - n\mu}{\sigma\sqrt{n}} = \frac{S_n - m_n - n\mu}{\sigma\sqrt{n}}.$$

Since m_n converges to a finite random variable for $\mu > 0$, the result (i) follows from the central limit theorem for MRW (see Theorem 5.1). The result (ii) follows in a similar manner. □

Remark 7.1. In the case where $\sigma = \infty$, one may expect the limit distribution of S_n to be in the domain of attraction of a stable law. The results given by Heyde [44] can then be invoked to obtain the limit distribution of the waiting time.

7.4 Markov-modulated M/M/1 Queue

In this section we consider a Markov-modulated M/M/1 queueing system where the interarrival time u_n and the service time v_n have the following joint exponential density given J_n and J_{n-1}. i.e.

$$P\{u_n \le x, v_n \le y | J_{n-1} = j, J_n = k\} = (1 - e^{-\lambda_{jk}x})(1 - e^{-\mu_{jk}y}) \quad (7.28)$$

for $x, y \ge 0$. The resulting transition measure of the associated MRW is given by

$$Q_{jk}\{dx\} = \begin{cases} \frac{p_{jk}\lambda_{jk}\mu_{jk}}{\lambda_{jk}+\mu_{jk}} e^{\lambda_{jk}x} & \text{if } x \le 0 \\ \frac{p_{jk}\lambda_{jk}\mu_{jk}}{\lambda_{jk}+\mu_{jk}} e^{-\mu_{jk}x} & \text{if } x \ge 0. \end{cases} \quad (7.29)$$

We consider the following four conditions.

C_1: Arrival rate of the next customer does not depend on the present customer-type (the J-state); i.e. $\lambda_{jk} = \lambda_k$, $\forall j$.

C_2: Service rate of the present customer does not depend on the next customer-type, i.e. $\mu_{jk} = \mu_j$, $\forall k$.

C_3: Arrival rate of the next customer only depends on the present J-state; i.e. $\lambda_{jk} = \lambda_j$, $\forall k$

C_4: Service rate of the present customer only depends on the next customer-type, i.e. $\mu_{jk} = \mu_k$, $\forall j$.

When all these conditions hold it reduces to the classical M/M/1 system. Conditions C_1 and C_2 are more often met in most manufacturing systems as they imply that the arrival rate of a job to a machine is independent of the preceding one and that the service rate of a job under process is independent of its successor. Condition C_3 implies that the arrival rate of the job depends only on the preceding one and condition C_4 implies that the rate of service depends on the next job to be done.

Letting

$$\mathbf{A} = (a_{jk}) = \left(\frac{p_{jk}\lambda_{jk}}{\mu_j + \lambda_{jk}} \Big/ \sum_i \frac{p_{ji}\lambda_{ji}}{\mu_j + \lambda_{ji}} \right) \quad (7.30)$$

$$\mathbf{B} = (b_{kj}) = \left(\frac{\pi_k p_{kj}\mu_{kj}}{\mu_{kj} + \lambda_j} \Big/ \sum_i \frac{\pi_i p_{ij}\mu_{ij}}{\mu_{ij} + \lambda_j} \right) \quad (7.31)$$

$$\mathbf{C} = (c_{jk}) = \left(\frac{p_{jk}\mu_{jk}}{\lambda_j + \mu_{jk}} \Big/ \sum_i \frac{p_{ji}\mu_{ji}}{\lambda_j + \mu_{ji}} \right) \quad (7.32)$$

$$\mathbf{D} = (d_{kj}) = \left(\frac{\pi_k p_{kj}\lambda_{kj}}{\lambda_{kj} + \mu_j} \Big/ \sum_i \frac{\pi_i p_{ij}\lambda_{ij}}{\lambda_{ij} + \mu_j} \right) \quad (7.33)$$

we have the following partial lack of memory properties.

Lemma 7.1. (Partial lack of memory property)
(a). *Under condition C_2, for all $x > 0$ and $y \geq 0$:*

$$P_j\{X_1 \leq x + y, J_1 = k | X_1 > y\} = a_{jk}[1 - \exp(-\mu_j x)]. \tag{7.34}$$

(a). *Under condition C_1, for all $x \leq 0$ and $y > 0$:*

$$P_j\{\widehat{X}_1 \geq x - y, \widehat{J}_1 = k | \widehat{X}_1 \leq -y\} = b_{jk}[1 - \exp(\lambda_j x)]. \tag{7.35}$$

(c). *Under condition C_3, for all $x \leq 0$ and $y > 0$:*

$$P_j\{X_1 \geq x - y, J_1 = k | X_1 \leq -y\} = c_{jk}[1 - \exp(\lambda_j x)]. \tag{7.36}$$

(d). *Under condition C_4, for all $x > 0$ and $y \geq 0$:*

$$P_j\{\widehat{X}_1 \leq x + y, \widehat{J}_1 = k | \widehat{X}_1 > y\} = d_{jk}[1 - \exp(-\mu_j x)]. \tag{7.37}$$

Proof. (a). For all $x, y \geq 0$, we have

$$P_j\{X_1 \leq x + y, J_1 = k | X_1 > y\} = \frac{P_j\{y < X_1 \leq x + y, J_1 = k\}}{P_j\{X_1 > y\}}$$
$$= \frac{p_{jk}\lambda_{jk}}{\mu_{jk} + \lambda_{jk}} \left[e^{-\mu_{jk} y} - e^{-\mu_{jk}(x+y)}\right] \Big/ \sum_i \frac{p_{ji}\lambda_{ji}}{\lambda_{ji} + \mu_{ji}} e^{-\mu_{ji} y}$$

which reduces to (7.34) by (7.30) and condition C_2.
The proofs for (b), (c), and (d) are similar. □

Theorem 7.5. *For the Markov-modulated M/M/1 queue we have:*

$$\chi^+(z, \omega) = \eta(z) \cdot \Theta_a(\omega) \quad \text{if } C_2 \text{ is satisfied} \tag{7.38}$$
$$\widehat{\chi}^-(z, \omega) = \widehat{\xi}(z) \cdot \Theta_b(\omega) \quad \text{if } C_1 \text{ is satisfied} \tag{7.39}$$
$$\chi^-(s, \omega) = \xi(z) \cdot \Theta_c(\omega) \quad \text{if } C_3 \text{ is satisfied} \tag{7.40}$$
$$\widehat{\chi}^+(z, \omega) = \widehat{\eta}(z) \cdot \Theta_d(\omega) \quad \text{if } C_4 \text{ is satisfied} \tag{7.41}$$

where

$$\eta(z) = (E_j[z^T; J_{T-1} = k]), \qquad \xi(z) = (E_j[z^{\bar{T}}; J_{\bar{T}-1} = k]), \tag{7.42}$$
$$\Theta_a(\omega) = \left(a_{jk}\frac{\mu_j}{\mu_j - i\omega}\right), \qquad \Theta_b(\omega) = \left(b_{jk}\frac{\lambda_j}{\lambda_j + i\omega}\right), \tag{7.43}$$
$$\Theta_c(\omega) = \left(c_{jk}\frac{\lambda_j}{\lambda_j + i\omega}\right), \qquad \Theta_d(\omega) = \left(d_{jk}\frac{\mu_j}{\mu_j - i\omega}\right). \tag{7.44}$$

Proof. Let

$$f_{jk}^{(n)}(x) = P_j\{T = n, 0 < S_T \le x, J_T = k\}.$$

Then, we have

$$f_{jk}^{(n)}(x) = \int_{-\infty}^{0} \sum_{l \in E} P_j\{T = n, S_{n-1} \in dy, J_{n-1} = l\} \cdot P_l\{X_1 \le x + y, J_1 = k | X_1 > y\}.$$

From (7.34) we know that

$$P_l\{X_1 \le x + y, J_1 = k | X_1 > y\} = a_{lk}[1 - \exp(-\mu_l x)]$$

which is independent of y. So we have

$$f_{jk}^{(n)}(x) = \sum_{l \in E} P_j\{T = n, J_{n-1} = l\} a_{lk}[1 - \exp(-\mu_l x)].$$

Now

$$\begin{aligned}
\chi_{jk}^{+}(z, \omega) = E_j[z^T e^{i\omega S_T}; J_T = k] &= \int_0^{\infty} \sum_{n=1}^{\infty} z^n e^{i\omega x} f_{jk}^{(n)}(dx) \\
&= \sum_{l \in E} E_j[z^T; J_{T-1} = l] \int_0^{\infty} a_{lk}\mu_l \exp[(-\mu_l + i\omega)x] dx \\
&= \sum_{l \in E} \eta_{jl}(z) \, [\Theta_a(\omega)]_{lk}
\end{aligned}$$

which proves (7.38). The proofs for (7.39)–(7.41) are similar. □

Equation (7.40) shows that the number of customers served in a busy-period is conditionally independent of the duration of the idle period that follows it, given the last customer-type served. Moreover, we have the following result.

Lemma 7.2. *Under condition C_3 the transition density of the idle period given the last customer type served in an busy-period is given by*

$$H_{jk}\{dx\} = c_{jk} \lambda_j exp(-\lambda_j x) \tag{7.45}$$

i.e. given that the last customer type served in an busy-period is j, the length of the idle period is exponentially distributed with rate λ_j.

Proof. First we note that $I_1 = -S_{\bar{T}}$. From (7.44) we see that $[\Theta_c(\omega)]_{jk}$ is the characteristic function for the random variable having the density function

$$c_{jk} \lambda_j \exp(\lambda_j x).$$

The result follows by changing the sign and the assertion follows from the fact that the row sum of c_{jk} is 1. □

Next, for the purpose of evaluating η and $\widehat{\xi}$, we state the following matrix identities which hold under conditions C_1 and C_2. Similar identities involving $\widehat{\eta}$ and ξ can be obtained under conditions C_3 and C_4.

Proposition 7.3. *Under conditions C_1 and C_2, η and $\widehat{\xi}$ satisfy the following matrix equations.*

$$z\mathbf{P} = \mathbf{B}'^{-1}\mathbf{D}_\pi^{-1}\widehat{\xi}'\mathbf{D}_\pi + \eta\mathbf{A}^{-1} - \mathbf{B}'^{-1}\mathbf{D}_\pi^{-1}\widehat{\xi}'\mathbf{D}_\pi\,\eta\mathbf{A}^{-1} \tag{7.46}$$

$$\begin{aligned} z\Im = {} & \mathbf{B}'^{-1}\mathbf{D}_\pi^{-1}\mathbf{L}\widehat{\xi}'\mathbf{D}_\pi - \eta\mathbf{M}\mathbf{A}^{-1} - \mathbf{B}'^{-1}\mathbf{D}_\pi^{-1}\mathbf{L}\widehat{\xi}'\mathbf{D}_\pi\eta\mathbf{A}^{-1} \\ & + \mathbf{B}'^{-1}\mathbf{D}_\pi^{-1}\,\widehat{\xi}'\mathbf{D}_\pi\,\eta\mathbf{M}\mathbf{A}^{-1} \end{aligned} \tag{7.47}$$

where

$$\Im_{jk} = \frac{p_{jk}(\lambda_j - \mu_k)}{\lambda_j\mu_k}, \qquad \mathbf{M} = diag(\mu_j^{-1}), \qquad \mathbf{L} = diag(\lambda_j^{-1}). \tag{7.48}$$

Proof. From (7.19) we have

$$\mathbf{I} - z\Phi(\omega) = \mathbf{D}_\pi^{-1}[\mathbf{I} - \widehat{\xi}(z)\Theta_b(\omega)]'\,\mathbf{D}_\pi[\mathbf{I} - \eta(z)\Theta_a(\omega)].$$

On the left side of this identity we have

$$\begin{aligned} \delta_{jk} - z & \frac{p_{jk}\lambda_j\mu_k}{(\lambda_j + i\omega)(\mu_k - i\omega)} \\ & = \delta_{jk} - z\left[\frac{p_{jk}\lambda_j\mu_k(\lambda_j\mu_k + \omega^2)}{(\lambda_j^2 + \omega^2)(\mu_k^2 + \omega^2)} + \frac{i\omega p_{jk}\lambda_j\mu_k(\lambda_j - \mu_k)}{(\lambda_j^2 + \omega^2)(\mu_k^2 + \omega^2)}\right] \end{aligned}$$

while on the right side we have

$$\begin{aligned} \delta_{jk} - & \left(\sum_m \frac{b_{jm}\lambda_m(\lambda_m - i\omega)}{\pi_j(\lambda_m^2 + \omega^2)}\widehat{\xi}'_{mk}\pi_k + \sum_m \eta_{jm}\frac{a_{mk}\mu_m(\mu_m + i\omega)}{\mu_m^2 + \omega^2}\right. \\ & \left. - \sum_m \left[\sum_q \frac{b_{jq}\lambda_q(\lambda_q - i\omega)}{\pi_j(\lambda_q^2 + \omega^2)}\widehat{\xi}'_{qm}\pi_k\right]\left[\sum_r \eta_{mr}\frac{a_{rk}\mu_r(\mu_r + i\omega)}{\mu_r^2 + \omega^2}\right]\right). \end{aligned}$$

For the real part we let $\omega \to 0$ on both sides. For the imaginary part we divide both sides by ω then we let $\omega \to 0$. The result then follows. □

Remark 7.2. In the case where $E = \{j\}$, we have from (7.46) and (7.47)

$$z = \eta + \xi - \eta\xi \quad \text{and} \quad \frac{z(\lambda - \mu)}{\lambda\mu} = \frac{\eta}{\mu} - \frac{\xi}{\lambda} - \frac{\eta\xi(\lambda - \mu)}{\lambda\mu}$$

which give the results

$$\xi = \frac{(\lambda + \mu) - \sqrt{(\lambda + \mu)^2 - 4z\mu\lambda}}{2\lambda}$$

and

$$\eta = \frac{(\lambda + \mu) - \sqrt{(\lambda + \mu)^2 - 4z\mu\lambda}}{2\mu}.$$

These agree with the known results for the classical M/M/1 system.

Chapter 8

A Storage Model for Data Communication Systems

In this chapter we analyze a storage process where the input as well as the output are modulated by an underlying Markov chain. The joint process of cumulative input, cumulative output, and the underlying Markov chain constitutes a Markov-additive process (MAP). Such processes arise frequently from models for data communication systems, which have received considerable attention in the literature.

8.1 Introduction

The specific model studied in this chapter was motivated by the pioneering paper by Anick, Mitra and Sondhi [2], who proposed the following model for a multiple source data-handling system. There are N sources of messages, which may be *on* or *off* from time to time. A switch receives these messages at a unit rate from each *on* source and transmits them at a fixed maximum rate c ($1 \leq N < \infty$, $0 < c < \infty$), storing the messages that it cannot transmit in a buffer of infinite capacity. The sources act independently, the durations of *on* and *off* times being independent random variables with exponential densities. Denoting by $J(t)$ the number of *on* sources at time t, these assumptions amount to the statement that $J = \{J(t),\, t \geq 0\}$ is a birth and death process on the state space $\{0, 1, \ldots, N\}$. Also, let us denote by $Z(t)$ the buffer content at time t. The model states that

$$Z(t) = Z(0) + \int_0^t z(Z(s), J(s))\, ds, \tag{8.1}$$

[8]This chapter is based on the article by N.U. Prabhu and A. Pacheco, 'A storage model for data communication systems', *Queueing Systems Theory Appl.* **19**, 1-2, pp. 1–40, 1995, and contents are reproduced with permission of the publisher, Springer.

where z is the net input rate given by

$$z(x,j) = \begin{cases} j-c & x>0 \\ (j-c)^+ & x=0 \end{cases}, \tag{8.2}$$

where $x^+ = \max(x,0)$. The authors derive the steady state distribution of the process (Z,J), employing matrix-algebraic techniques.

Gaver and Lehoczky [40] investigate a model for an integrated circuit and packet switched multiplexer in which there are two types of input – data and voice calls. The input of data occurs continuously at a constant rate c_1 (fluid input), and is transmitted at rate c_2. There are $s+u$ output channels, of which s are reserved for data transmission, while the remaining u are shared by data and voice calls, with calls having priority with preemptive discipline. Calls arrive at a Poisson rate and have independent holding times with exponential density. Calls that find all u channels that service them busy are lost. Data that cannot be transmitted are stored in a buffer of infinite capacity. Let $Z(t)$ be the buffer content and $J(t)$ the number of calls in progress at time t. Then the model states that J is identical with the number of customers in the Erlang loss system $M/M/u/u$, while $Z(t)$ satisfies Eq. (8.1) with the net input rate z given by

$$z(x,j) = \begin{cases} c_1 - c_2\,(s+u-j) & x>0 \\ [c_1 - c_2\,(s+u-j)]^+ & x=0 \end{cases}. \tag{8.3}$$

We note that in the two models described above the maximum output rate is constant. However, in practice, such factors as failures of output channels may induce over time a stochastic behavior on this rate. This is accounted for by Mitra [70], who investigates a fluid model for a system with two groups of machines connected by a buffer. The first group of machines (producers) produces a fluid which is transferred to the buffer and is consumed by a second group of machines (consumers). These machines are *in service* or *in repair* states from time to time and the assumptions for *in service* times and *in repair* times are similar to those on Anick, Mitra and Sondhi [2] for *on* and *off* periods. The buffer has capacity $C\ (0 < C \leq \infty)$, while the production rate and consumption rate by machine, assumed to be constants, are c_1 and c_2. We let $J_1(t)$ and $J_2(t)$ denote respectively the number of *producers* and *consumers* that are *in service* at time t, from the total of N *producers* and M *consumers*. It follows that $J_1 = \{J_1(t)\, t \geq 0\}$ and $J_2 = \{J_2(t)\, t \geq 0\}$ are independent Markov processes on $\{0,1,\dots,N\}$ and $\{0,1,\dots,M\}$, respectively. Again, if we denote by $Z(t)$ the buffer content at time t, the model states that

$$Z(t) = Z(0) + \int_0^t z(Z(s), J_1(s), J_2(s))\, ds, \tag{8.4}$$

where the net input rate z is given by

$$z(x, j_1, j_2) = \begin{cases} (c_1 j_1 - c_2 j_2)^+ & x = 0 \\ c_1 j_1 - c_2 j_2 & 0 < x < C \\ \min(c_1 j_1 - c_2 j_2, 0) & x = C \end{cases}. \tag{8.5}$$

This model is applicable in communication-integrated systems with *producers* being interpreted as sources and *consumers* as channels.

Van Doorn, Jagers and de Wit [114] consider a reservoir (dam) model regulated by a birth and death process J on the state space $\{0, 1, \dots, N\}$ with arbitrary birth and death rates. Their assumptions seem to imply the equation (8.1) for the dam content $Z(t)$, with the net input rate z given by

$$z(x, j) = \begin{cases} z_j & x > 0 \\ z_j^+ & x = 0 \end{cases}. \tag{8.6}$$

It should be noted that the net input rate of this model is essentially of the same type as (8.2) and (8.3). However, the underlying Markov chain makes a weaker presence in this model than in the context of data communication systems as described above.

A model for an access regulator was investigated by Elwalid and Mitra [33]. Here the input of data is regulated by a continuous time Markov chain J with a finite number of states, the input rate being $\lambda(j)$ when J is in state j (the so called Markov-modulated fluid source). In addition to a finite data buffer receiving messages (cells), as in Mitra [70], there is a token buffer also of finite size, receiving tokens continuously at rate r. In order to get transmitted, cells have to be combined with tokens. Some cells are lost because of a regulator that controls transmission subject to a peak rate ν.

Models similar to those described here arise also in manufacturing systems. Browne and Zipkin [23] consider an (s, S) inventory system where the demand rate is controlled by an underlying Markov chain J. The model states that the demand in $(0, t]$ is given by

$$D(t) = \int_0^t d(J(s))\, ds. \tag{8.7}$$

The authors study steady state properties of the system. Sevast'yanov [105] analyzes a system with N machines in tandem with buffers in between. Each item has to be processed sucessively by machines $1, 2, \dots, N$, and items that cannot be processed by a machine are stored in the preceding buffer. The system may be viewed as a tandem storage model modulated by an auxiliary process J indicating the number of machines in operation. With appropriate assumptions, J will turn to be an N-dimensional Markov chain, each component of which indicates whether or not the corresponding machine is in operation. This model and the one by Elwalid and Mitra [33] are simple examples of storage networks modulated by an underlying Markov chain.

The model investigated in this chapter is an extension of the model of Anick, Mitra and Sondhi [2]. We first note that (8.1) can be written as

$$Z(t) = Z(0) + X(t) - \int_0^t r(Z(s), J(s))\, ds, \tag{8.8}$$

with

$$X(t) = \int_0^t J(s)\, ds \tag{8.9}$$

and

$$r(x, j) = \begin{cases} c & x > 0 \\ \min(c, j) & x = 0 \end{cases}. \tag{8.10}$$

Here $X(t)$ is the total input into the buffer in the time-interval $(0, t]$ and r is the release rate. If we denote by $(X, J) = \{(X, J)(t),\, t \geq 0\}$ the input process, then it is easily seen that (X, J) is an MAP with the additive component X consisting of a nonnegative drift that depends on the current state of the Markov chain.

In our model we denote by J a Markov chain on a countable state space and assume that the input process (X, J) is an MAP, where, in addition to a nonnegative drift $a(j)$ when J is in state j, X has nonnegative jumps whose rate and size depend on the state of J. This formulation provides for the possibility of two sources of input, one slow source bringing in data in a fluid fashion and the other bringing in packets. As for the demand for transmission of data we assume that it arises at a rate $d(j)$ when the current state of J is j (similar to what happens in the model considered by Mitra [70]). The storage policy is to meet the demand if physically possible, i.e. the output channel is used the most possible. We postulate that the buffer content $Z(t)$ at time t satisfies the integral equation (8.8) with our input process (X, J) and the release rate r given by

$$r(x, j) = \begin{cases} d(j) & x > 0 \\ \min(d(j), a(j)) & x \leq 0 \end{cases}. \tag{8.11}$$

We prefer, however, to write (8.8) slightly differently, as follows. Let X_0 be the part of the input X free of drift, so that

$$X(t) = X_0(t) + \int_0^t a(J(s))\, ds. \tag{8.12}$$

Then (8.8) becomes

$$Z(t) = Z(0) + X_0(t) + \int_0^t z(Z(s), J(s))\, ds, \tag{8.13}$$

with

$$z(x,j) = \begin{cases} a(j) - d(j) & x > 0 \\ [a(j) - d(j)]^+ & x \leq 0 \end{cases} \tag{8.14}$$

At this stage we make an additional assumption, namely that the demand rate exceeds the input rate from the slow source, so that $d(j) > a(j)$. Thus, finally we obtain

$$Z(t) = Z(0) + X_0(t) - \int_0^t \mathbf{1}_{\{Z(s)>0\}}\, d_1(J(s))\, ds, \tag{8.15}$$

where $d_1(j) = d(j) - a(j) > 0$. Accordingly, we may accept (8.15) as our postulate, changing X_0 to X (free of drift) and d_1 to d.

The chapter is organized as follows. In Sec. 8.2 we introduce our model and prove the existence and uniqueness of the solution of the basic integral equation. In Sec. 8.3 we investigate the properties of the actual and unsatisfied demands, and in Sec. 8.4 the inverse of the actual demand. If we denote by $D(t)$ the total demand in $(0,t]$ and by $B(x)$ its inverse, then the main result is that $\{(D,J)(t),\, t \geq 0\}$ and $\{(B, J \circ B)(x),\, x \geq 0\}$ are both MAPs consisting of just a drift. The busy-period problem is investigated in Sec. 8.5. If T denotes the busy-period, it follows that $(T, J \circ T)$ is an MAP with both drift and jumps. Sec. 8.6 contains some additional notations and results of a technical nature. Transforms of the various processes of interest are obtained in sections 8.7-8.9. The steady state behavior of the model is investigated in Sec. 8.10.

8.2 The Model

(a) *The Input.* The input $\{X(t),\, t \geq 0\}$ is such that $(X,J) = \{(X,J)(t),\, t \geq 0\}$ is a Markov-compound Poisson process (MCPP) with nonnegative jumps and no drift, and infinitesimal generator given by

$$\mathcal{A}_{(X,J)}\, f(x,j) = \sum_{k \in E} \sigma_{jk} \int_{0-}^{\infty} [f(x+y,k) - f(x,j)] M_{jk}\{dy\}, \tag{8.16}$$

where E is a countable set, and $f(x,j)$ is a bounded function on $\mathbf{R}_+ \times E$ such that for each fixed j, f is continuous and has a bounded continuous derivative $\frac{\partial f}{\partial x}(x, j)$. Thus, jumps in X associated with a transition from state j to state k in J occur at rate σ_{jk} and have distribution M_{jk}. The distributions M_{jk}, $j, k \in E$, are concentrated on $\mathbf{R}_+$.

(b) *The Demand.* The demand is Markov linear, and a function of the same continuous-time Markov chain J as the input, namely

$$D(t) = \int_0^t d(J(s))\, ds, \tag{8.17}$$

where d is a positive measurable function. We define d_- and d_+ as follows:

$$0 \le d_- = \inf_{j \in E} d(j) \le \sup_{j \in E} d(j) = d_+ \le +\infty. \tag{8.18}$$

(c) *The Release Policy.* The release rule states that the demand is satisfied if physically possible so that the amount of unsatisfied demand is given by (see Remark 8.1 and Example 8.1 below)

$$I(t) = \int_0^t d(J(s))\, \mathbf{1}_{\{Z(s) \le 0\}}\, ds. \tag{8.19}$$

The *net input* process is given by $Y(t) = X(t) - D(t)$, and the *output* process is given by $O(t) = D(t) - I(t)$. The underlying Markov chain J, which modulates the jumps on the input as well as the demand rates has, from (8.16), infinitesimal generator

$$\mathcal{A}_J f(j) = \lambda_j \sum_{k \in E} [f(k) - f(j)]\, p_{jk}, \qquad j \in E, \tag{8.20}$$

where

$$\Sigma = (\sigma_{jk}) = \Lambda P = (\lambda_j\, \delta_{jk})\, (p_{jk}), \tag{8.21}$$

with P being a stochastic matrix. We call Σ the *transition rate matrix*, the diagonal matrix Λ the *intensity matrix*, and P the *transition probability matrix* of J. We assume J is non-explosive, which holds, in particular, if $\sup \lambda_j < \infty$. In case J is ergodic, we denote by the row vector $\pi = \{\pi_j,\, j \in E\}$ its stationary distribution and let $\Pi = \mathbf{e}\pi$, where $\mathbf{e}$ is a column vector of ones, so that

$$\Pi = \left(\lim_{t \to \infty} P[J(t) = k \mid J(0) = j] \right).$$

We let $\tilde{J} = \{J_n,\, n \ge 0\}$ be the associated discrete time Markov chain (DTMC) with stationary transition probability matrix P. We may then consider $J_n = J(T_{n+1}^-)$, $n \ge 0$, with

$$T_n = \text{ time at which the } n\text{-th transition in } J \text{ occurs.}$$

Having defined the storage process, one very important question we need to address is of the following type:

What are the sufficient conditions for the storage process to be a friendly object, e.g. measurable with respect to the σ-field in which we consider the processes X and D; progressively-measurable with respect to a given filtration of the process (X, D), etc.?

Theorem 8.1. *We are given a stochastic process (X, J), as defined above, on a probability space $(\Omega, \mathcal{F}, \mathcal{P})$. Consider the integral equation*

$$Z(t) = Z(0) + X(t) - D(t) + \int_0^t d(J(s))\, \mathbf{1}_{\{Z(s)\leq 0\}}\, ds, \tag{8.22}$$

where $Z(0) \geq 0$. We have ($\mathcal{P}$-a.s.):

(a) *The integral Eq. (8.22) has the unique solution*

$$Z(t) = Z(0) + Y(t) + I(t)\,, \tag{8.23}$$

where

$$I(t) = [-m(t) - Z(0)]^+, \tag{8.24}$$

with

$$m(t) = \inf_{0\leq\tau\leq t} Y(\tau). \tag{8.25}$$

(b) *D and Y are $\mathcal{F}$-measurable stochastic processes, and Z and I are $(\mathcal{F} \bigvee \sigma(Z(0)))$-measurable.*
(c) *D and I have continuous sample paths, and Y and Z have* càdlàg *(right-continuous with left-hand limits) sample paths with all jumps being positive.*
(d) *If, in addition, we are given a filtration $\{\mathcal{F}_t, t \geq 0\}$ and (X, J) is adapted to the filtration $\{\mathcal{F}_t\}$ then, in addition to the results of* (a)–(c), *J, X, D and Y are progressively measurable with respect to $\{\mathcal{F}_t\}$, and I and Z are progressively measurable with respect to $\{\mathcal{F}_t \bigvee \sigma(Z(0))\}$.*
(e) *If $\mathcal{F}^o_t = \sigma\{(X, J)(s), s \leq t\}$ and $\mathcal{F}_t$ is the proper completion of $\mathcal{F}^o_t$, then D and Y are progressively measurable with respect to $\{\mathcal{F}_t\}$, and I and Z are progressively measurable with respect to the proper completion of $\{\mathcal{F}^o_t \bigvee \sigma(Z(0))\}$.*

Proof. During the proof, it should be kept in mind that $\mathcal{P}$-a.s.: *(i)* (X, J) has *càdlàg* sample paths, *(ii)* in a finite time interval, the number of jumps of (X, J) is finite (since J is non-explosive), and *(iii)* all jumps of X are nonnegative. For simplicity we omit $\mathcal{P}$-a.s. in the proof.

(a). Let ω, a point in the sample space satisfying *(i)-(iii)*, and $t > 0$ be fixed. From the setting of the theorem and Eq. (8.22) we easily see that $Z(t) \geq 0, \forall t$, and for $\tau \in [0,t]$, with $Z(0^-) = Z(0)$,

$$\begin{aligned} Z(t) &\geq Z(t) \ - Z(\tau^-) \\ &= Y(t) - Y(\tau^-) + \int_\tau^t d(J(s))\, \mathbf{1}_{\{Z(s)\leq 0\}}\, ds \\ &\geq Y(t) \ - Y(\tau^-). \end{aligned} \tag{8.26}$$

If $Z(\tau)$ is never zero for $\tau \in (0,t)$, then, from (8.22) and (8.26),

$$Z(t) \ = Z(0) + Y(t) \geq \sup_{0\leq\tau\leq t} [Y(t) \ - Y(\tau^-)]; \tag{8.27}$$

otherwise let $t_0 = \max\{\tau : 0 < \tau \leq t, Z(\tau^-) = 0\}$ and use (8.26) and (8.22) to get

$$Z(t) \geq \sup_{0\leq\tau\leq t} [Y(t) \ - Y(\tau^-)] \geq [Y(t) \ - Y(t_0^-)] = Z(t). \tag{8.28}$$

Since $Z(t) \geq Z(0) + Y(t), \forall t \geq 0$, using (8.27) and (8.28), it follows easily that

$$\begin{aligned} Z(t) &= \max\{Z(0) + Y(t), \sup_{0\leq\tau\leq t} [Y(t) - Y(\tau-)]\} \\ &= \max\{Z(0) + Y(t), Y(t) - m(t)\}, \end{aligned} \tag{8.29}$$

which, with (8.22), implies that

$$\begin{aligned} I(t) &= \int_0^t d(J(s))\, \mathbf{1}_{\{Z(s)\leq 0\}}\, ds \\ &= Z(t) - Z(0) - Y(t) = \max\{0, -m(t) - Z(0)\}, \end{aligned}$$

as required.

(b). Since (X, J) is right-continuous it is also $\mathcal{F}$-measurable by Proposition 1.13 and Remark 1.14 in Karatzas ([50], Chapter 1). The $\mathcal{F}$-measurability of J implies the $\mathcal{F}$-measurability of $d(J(t))$, which in turn implies the $\mathcal{F}$-measurability of D by Remark 4.6(i) in Karatzas ([50], Chapter 1), and therefore $Y = X - D$ is also $\mathcal{F}$-measurable, as does $m = \{m(t), t \geq 0\}$ given by (8.25). Since m is also $(\mathcal{F} \bigvee \sigma(Z(0)))$-measurable so are I, by (8.24), and Z, by (8.23).

(c). Since J has a finite number of jumps in a finite interval, the continuity of the sample paths of D and I follows. Since X has *càdlàg* non-decreasing sample paths and D has continuous sample paths, $Y = X - D$ has *càdlàg* sample paths with all jumps being positive; this and the fact that I has continuous sample paths imply, by (8.23), that Z has *càdlàg* sample paths with all jumps being positive.

(d). The proof follows along the same lines of (b); the fact that (X, J) is progressively measurable with respect to $\{\mathcal{F}_t\}$ follows from the fact that it is adapted to $\{\mathcal{F}_t\}$ and has *càdlàg* sample functions, by Proposition 1.13 in Karatzas ([50], Chapter 1).

(e). The statement follows from (d). □

Remark 8.1. It follows from Theorem 8.1 that

$$I(t) = \int_0^t d(J(s))\, \mathbf{1}_{\{Z(s)=0\}}\, ds, \tag{8.30}$$

since $Z(t) \geq 0$, $\forall t \geq 0$. If we replace Eq. (8.22) by

$$Z(t) \; = Z(0) + X(t) \; - D(t) \; + \int_0^t d(J(s))\, \mathbf{1}_{\{Z(s)=0\}}\, ds, \tag{8.31}$$

(8.23) and (8.24) is also a solution of (8.31). However, (8.31) may not have an unique solution as we show in the following example.

Example 8.1. Consider $E = \{1\}$, $\Sigma = (0)$, $M_{11}\{0\} = 1$, $d(1) = 1$ and $Z(0) = 0$ a.s.. Note that $X(t) = 0$ and $D(t) = t$, a.s.. Trivially, $Z(t) = -t$ and $I(t) = 0$ is a solution of (8.31) (in addition to the solution $Z(t) = 0$ and $I(t) = t$, correspondent to (8.23) and (8.24)).

8.3 The Actual and the Unsatisfied Demands

We recall the definition of an additive functional of a Markov process (see [17]).

Definition 8.1. Let $\{R(t),\, t \geq 0\}$ be a Markov process, $\mathcal{F}_t^o = \sigma\{R(s), s \leq t\}$ and $\mathcal{F}_t$ the proper completion of $\mathcal{F}_t^o$. Then $\{\mathcal{F}_t, t \geq 0\}$ is a right-continuous filtration for $\{R(t),\, t \geq 0\}$.

A family $\{A(t),\, t \geq 0\}$ of functions with values in $[0, \infty]$ is called an additive functional of $\{R(t),\, t \geq 0\}$ provided:

(1) $t \to A(t)$, $t \geq 0$, is non-decreasing, right-continuous, and $A(0) = 0$, a.s. for all initial distributions.
(2) For each $t, A(t) \in \mathcal{F}_t$.
(3) For each $s, t \geq 0, A(t + s) = A(t) + A(s) \circ \phi_t$, a.s. for any initial distribution, where ϕ is a shift operator,

$$\phi_t : \Omega \to \Omega \text{ such that } (\phi_t(\omega))(r) = \omega(r + t),\ r \geq 0.$$

Remark 8.2. With the setting of Definition 8.1, a functional A of $\{R(t),\, t \geq 0\}$ of the form

$$A(t) = \int_0^t f(R(s))\, ds, \tag{8.32}$$

where f is a bounded non-negative measurable function, is called a *classical* additive functional by Blumenthal [17], but as remarked in Blumenthal ([18], p. 151), A is still an additive functional of R even when f is not bounded provided $A(t)$ is finite for all t a.s.

Lemma 8.1. *The demand process $\{D(t),\, t \geq 0\}$ has the following properties:*

(a) *D is a continuous additive functional of the Markov chain J.*
(b) *D has strictly increasing sample paths, for all initial distributions.*
(c) *If $d_- > 0$, then $D(\infty) = \lim_{t\to\infty} D(t) = \infty$, for all initial distributions.*
(d) *$d_- \leq \frac{D(t)}{t} \leq d_+$, for all $t > 0$.*
(e) *If J is ergodic, then $\lim_{t\to\infty} \frac{D(t)}{t} = \sum_{j\in E} d(j)\, \pi_j$, a.s.*
(f) *If J has a persistent state, then $\lim_{t\to\infty} D(t) = \infty$, a.s.*

Proof. (a). Since J is non-explosive, J has a finite number of transitions in finite time a.s. and this implies that $D(t) < \infty$ a.s., for all $t \geq 0$; the statement now follows from Remark 8.2.
(b). Since $d(j) > 0, \forall j \in E$, then all sample paths of $\{D(t),\, t \geq 0\}$ are strictly increasing regardless of the initial distribution and the result follows.
(c). The statement follows since, if $d_- > 0$, for any $t \geq 0$ and any initial distribution,

$$D(t) = \int_0^t d(J(s))\, ds \geq \int_0^t d_-\ ds = d_-\ t.$$

(d). Since for all $t > 0$, $d_- t \leq D(t) \leq d_+ t$, the statement follows.
(e). d is a positive measurable function on E, so that, if J is ergodic,

$$\begin{aligned}\frac{D(t)}{t} &= \frac{1}{t}\int_0^t d(J(s))\, ds = \frac{1}{t}\int_0^t \sum_{j\in E} d(j)\, \mathbf{1}_{\{J(s)=j\}}\, ds \\ &= \sum_{j\in E} \frac{d(j)}{t} \int_0^t \mathbf{1}_{\{J(s)=j\}}\, ds \to \sum_{j\in E} d(j)\, \pi_j, \text{ as } t \to \infty, \text{ a.s.}\end{aligned}$$

(f). Given a persistent state $j \in E$ (and since $d(j) > 0$),

$$D(t) = \int_0^t d(J(s))\, ds \geq \int_0^t d(j)\, \mathbf{1}_{\{J(s)=j\}}\, ds \to \infty, \text{ as } t \to \infty, \text{ a.s.},$$

and the statement follows. □

The demand process can be identified as a random clock, its speed at t being $d(J(t))$, and with transformed time scale $[0, D(\infty))$. Lemma 8.1 (c) and (f) gives sufficient conditions for the transformed time scale to be infinite a.s.; we note that the condition $d(j) > 0\,(j \in E)$, is not a sufficient one (in general) as the following example shows.

Example 8.2. Suppose $E = \{1, 2, 3, \ldots\}$, $d(j) = j^{-2}$, and Σ, the intensity matrix of J, is such that $\sigma_{j,j+1} = 1$, for $j = 1, 2, \ldots$, and $\sigma_{jk} = 0$ otherwise. Then, for any initial distribution,

$$E(D(\infty)) \leq \sum_{j=1}^{\infty} j^{-2} < \infty,$$

so that $D(\infty) < \infty$ a.s.

Since D is an additive functional of J, the process (D, J) is clearly an MAP and so is (Y, J). In Theorem 8.2 we give the infinitesimal generators of these two processes; in fact we are able to conclude that they are Markov-compound Poisson processes (MCPPs).

Theorem 8.2. *The processes* (D, J) *and* (Y, J) *are MCPPs; their infinitesimal generators are given by:*

$$\mathcal{A}_{(D,J)}\, f(x,j) = d(j)\frac{\partial}{\partial x} f(x,j) + \sum_{k \neq j} \sigma_{jk}[f(x,k) - f(x,j)], \tag{8.33}$$

$$\begin{aligned} \mathcal{A}_{(Y,J)}\, f(x,j) = &- d(j)\frac{\partial}{\partial x} f(x,j) \\ &+ \sum_{k \in E} \sigma_{jk} \int_{0-}^{\infty} [f(x+y,k) - f(x,j)] M_{jk}\{dy\}. \end{aligned} \tag{8.34}$$

Proof. For a Borel set A on $[0, +\infty]$ and $j, k \in E$, let

$$P_{jk}(A; t) = P\{D(t) \in A, J(t) = k \mid J(0) = j\}.$$

Then,

$$\begin{aligned} P_{jk}(A; h) = &\,\delta_{jk}(1 - \lambda_j h)\, \mathbf{1}_{\{h\, d(j) \in A\}} \\ &+ \int_0^h \sigma_{jk}\, e^{-\lambda_j \tau}\, e^{-\lambda_k (h-\tau)}\, \mathbf{1}_{\{\tau d(j) + (h-\tau) d(k) \in A\}}\, d\tau + o(h), \end{aligned}$$

and, consequently,

$$
\begin{aligned}
\mathcal{A}_{(D,J)} f(x,j) &= \lim_{h\to 0+}\left[\frac{1}{h}\sum_{k\in E}\int_{0-}^{\infty}[f(x+y,k)-f(x,j)]\,P_{jk}(\{dy\};h)\right] \\
&= d(j)\lim_{h\to 0+}\left[\int_{0-}^{\infty}\frac{f(x+y,j)-f(x,j)}{h\,d(j)}\,(1-\lambda_j\,h)\,\mathbf{1}_{\{h\,d(j)\,\in\,dy\}}\right] \\
&+\sum_{k\in E}\int_{0-}^{\infty}[f(x+y,k)-f(x,j)]\,\sigma_{jk} \\
&\qquad \lim_{h\to 0+}\left[\frac{e^{-\lambda_k h}}{h}\int_{0-}^{h}e^{-(\lambda_j-\lambda_k)\tau}\,d\tau\,\mathbf{1}_{\{\tau d(j)+(h-\tau)d(k)\,\in\,dy\}}\right] \\
&= d(j)\frac{\partial}{\partial x}f(x,j)+\sum_{k\neq j}\sigma_{jk}[f(x,k)-f(x,j)].
\end{aligned}
$$

This shows that (D,J) is an MCPP with drift function $d(j)$, $j\in E$, and no jumps. Thus, (Y,J) is an MCPP with drift function $-d(j)$, $j\in E$, and the same jumps as (X,J) so that, by (8.16), the infinitesimal generator of (Y,J) is as given in (8.34). □

We give in Theorem 8.3 some properties of the unsatisfied demand and of the infimum of the net input, which is naturally related to the unsatisfied demand by Theorem 8.1.

Theorem 8.3. *The unsatisfied demand $I=\{I(t),\,t\geq 0\}$ is a continuous additive functional of (Z,J); moreover, if $Z(0)=0$ a.s., then I is a continuous additive functional of (X,J).*

The negative of the infimum of the net input process, $-m$ (where m is given by (8.25)), is a continuous additive functional of (X,J).

Proof. Since (Z,J) is a Markov process, the fact that the unsatisfied demand I is a continuous additive functional of (Z,J) follows from the definition of the unsatisfied demand, as given in (8.19), and Remark 8.2.

Suppose now $Z(0)=0$ a.s., then (since $X(0)=0$ a.s.) $\sigma\{(X,J)(s),s\leq t\}$ is equal to $\sigma\{(Z,J)(s),s\leq t\}$ and I becomes a continuous additive functional of (X,J). The rest of the theorem follows trivially since, by Theorem 8.1, $I(t)=-m(t)$ a.s. in case $Z(0)=0$ a.s. □

8.4 The Inverse of the Demand

It is of interest to study the right-continuous inverse of the demand process $D(t)$, i.e., for $x \geq 0$ we want to know the amount of time needed for the demand to exceed x viewed as a stochastic process. Denoting this process as B we thus have

$$B(x) = \inf\{t : D(t) > x\}. \tag{8.35}$$

We study below some properties of the right-continuous inverse of an additive functional of a Markov process. The following result for inverses of additive functionals is partially mentioned by Blumenthal [17] but seems to call for a detailed proof.

Lemma 8.2. *If $\{A(t),\, t \geq 0\}$ is an additive functional of the Markov process $\{R(t),\, t \geq 0\}$ and $\tilde{A}(x)$ is its right-continuous inverse,*

$$\tilde{A}(x) = \inf\{t : A(t) > x\},$$

then:

(a) *$\tilde{A}$ has non-decreasing and a.s. right-continuous sample paths.*
(b) *If A has continuous and strictly increasing sample paths a.s., then a.s.:*
  (i) $\tilde{A}(x) = \infty$ for $x \in [A(\infty), \infty)$.
  (ii) $\tilde{A}(0) = 0$ and $\tilde{A}$ has continuous and strictly increasing sample paths on $[0, A(\infty))$.
(c) *For each $x, \tilde{A}(x)$ is a stopping time.*
(d) *For any $t \in \mathbf{R}_+$, a.s., $A(t) = \inf\{x : \tilde{A}(x) > t\}$.*

Proof. (a). If $x_1 \leq x_2$, then

$$\tilde{A}^*(x_2) = \{t : A(t) > x_2\} \subseteq \{t : A(t) > x_1\} = \tilde{A}^*(x_1),$$

so the indexed sets $\tilde{A}^*(x)$ are nonincreasing on x, and

$$\tilde{A}(x_1) = \inf \tilde{A}^*(x_1) \leq \inf \tilde{A}^*(x_2) = \tilde{A}(x_2).$$

This shows that $\tilde{A}(x)$ is non-decreasing. In the following we refer only to non-decreasing sample paths of A. Let $\{x_n\}$ be a decreasing sequence with limit $x \geq 0$; since $\{\tilde{A}(x_n)\}$ is a nonincreasing sequence bounded below by $\tilde{A}(x)$ the sequence converges in $[0, +\infty]$ and

$$\lim \tilde{A}(x_n) \geq \tilde{A}(x). \tag{8.36}$$

Now note that, for $y \geq 0$, $\tilde{A}^*(y)$ is either $(\tilde{A}(y), +\infty)$ or $[\tilde{A}(y), +\infty)$, so that

$$(\tilde{A}(y), +\infty) \subseteq \tilde{A}^*(y) \subseteq [\tilde{A}(y), +\infty), \forall y \geq 0.$$

Consequently,

$$(\tilde{A}(x), +\infty) \subseteq \tilde{A}^*(x) = \bigcup_n \tilde{A}^*(x_n) \subseteq \bigcup_n [\tilde{A}(x_n), +\infty) \subseteq [\lim \tilde{A}(x_n), +\infty),$$

which implies that $\lim \tilde{A}(x_n) \leq \tilde{A}(x)$. Thus, in view of (8.36), it follows that

$$\lim \tilde{A}(x_n) = \tilde{A}(x).$$

Therefore $\tilde{A}$ has right-continuous sample paths almost surely.
(b). We consider only continuous and strictly increasing sample paths of A such that $A(0) = 0$ (for which the associated sample paths of $\tilde{A}$ are non-decreasing and right-continuous) and keep the notation used in (a). If $A(\infty) < x < \infty$, then clearly $\tilde{A}(x) = \infty$ and, by right-continuity, also $\tilde{A}(A(\infty)) = \infty$; this proves *(i)*.

Trivially $\tilde{A}(0) = 0$ since $A(t) > 0$, $\forall t > 0$, thus to prove *(ii)* we need to show that $\tilde{A}$ is left-continuous and strictly increasing on $\mathcal{S} = [0, A(\infty))$. We have, $\tilde{A}^*(y) = (\tilde{A}(y), +\infty), \forall y \in \mathcal{S}$. Let $\{x_n\}$ be a non-negative increasing sequence with limit $x \in \mathcal{S}$, then $\{\tilde{A}(x_n)\}$ is a non-decreasing sequence bounded above by $\tilde{A}(x) < \infty$, so the sequence converges on $[0, +\infty)$ and

$$\lim \tilde{A}(x_n) \leq \tilde{A}(x). \tag{8.37}$$

Consequently

$$\begin{aligned}[\tilde{A}(x), +\infty) = \{t : A(t) \geq x\} &= \bigcap_n \{t : A(t) > x_n\} \\ &= \bigcap_n (\tilde{A}(x_n), +\infty) \supseteq (\lim \tilde{A}(x_n), +\infty),\end{aligned}$$

which implies that $\lim \tilde{A}(x_n) \geq \tilde{A}(x)$. Thus, in view of (8.37), it follows that

$$\lim \tilde{A}(x_n) = \tilde{A}(x).$$

This proves that $\tilde{A}(x)$ is left-continuous and thus continuous on $\mathcal{S}$. Suppose, with the same conditions, that $\tilde{A}(x)$ is not strictly increasing on $\mathcal{S}$, i.e.,

$$\exists y < x < A(\infty) \text{ such that } \tilde{A}(y) = \tilde{A}(x).$$

Then $(\tilde{A}(y), +\infty) = (\tilde{A}(x), +\infty)$ or, equivalently,

$$\{t : A(t) > y\} = \{t : A(t) > x\}.$$

Since $A(0) = 0$, this contradicts the fact that $A(t)$ is continuous and strictly increasing. Thus $\tilde{A}(x)$ is strictly increasing on $\mathcal{S}$. We have thus proved that $\tilde{A}$ has continuous and strictly increasing sample paths on $[0, A(\infty))$ a.s.

(c). Since for any $t, x \geq 0$

$$\{\tilde{A}(x) \leq t\} = \bigcap_n \{ \sup_{0 \leq \tau \leq t+\frac{1}{n}} A(\tau) > x\} \in \mathcal{F}_{t^+} = \mathcal{F}_t,$$

$\tilde{A}(x)$ is a stopping time.
(d). We prove the result for non-decreasing and right-continuous sample paths of A. For $t \geq 0$, let

$$\overline{A}(t) = \inf\{x : \tilde{A}(x) > t\},$$

then, using the fact that A and $\tilde{A}$ are non-decreasing and right-continuous and the definitions of $\tilde{A}$ and $\overline{A}$,

$$x < (>)\, \overline{A}(t) \Longrightarrow \tilde{A}(x) \leq (>)\, t \Longrightarrow x \leq (\geq)\, A(t).$$

This shows that $\overline{A}(t) = A(t)$, $\forall t \geq 0$, which proves the statement. □

Theorem 8.4. *For all initial distributions, the inverse of the demand has the following properties:*

(a) $B(0) = 0$ *and* B *has continuous and strictly increasing sample paths on* $[0, D(\infty))$ *and* $B(x) = \infty$ *for* $x \in [D(\infty), \infty)$.
(b) *For each* x, $B(x)$ *is a stopping time.*
(c) *For any* $t \in \mathbf{R}_+$,

$$D(t) = \inf\{x : B(x) > t\}. \tag{8.38}$$

(d) $\frac{1}{d_+} \leq \frac{B(x)}{x} \leq \frac{1}{d_-}$, *for all* $x > 0$.
(e) *If* J *is ergodic and* $d_+ < \infty$, *then* $\lim_{x\to\infty} \frac{B(x)}{x} = \left[\sum_{j\in E} d(j)\, \pi_j\right]^{-1}$, *a.s.*
(f) $\{(B, J \circ B)(x),\ 0 \leq x < D(\infty)\}$ *is an MAP with infinitesimal generator*

$$\mathcal{A}_{(B,J\circ B)}\, f(x,j) = \frac{1}{d(j)} \frac{\partial}{\partial x} f(x,j) + \sum_{k\neq j} \frac{\sigma_{jk}}{d(j)} \left[f(x,k) - f(x,j)\right]. \tag{8.39}$$

Proof. The statements (a),(b) and (c) follow in the same way as Lemma 8.2 (b)-(d) by Lemma 8.1 (a)-(b).
(d). It suffices to prove the result for $0 < d_- \leq d_+ < +\infty$; if so, given $x > 0$,

$$D(x/d_+) \leq \frac{x}{d_+}\, d_+ = x = \frac{x}{d_-}\, d_- \leq D(x/d_-).$$

Now, by the definition of $B(x)$, we have $\frac{x}{d_+} \leq B(x) \leq \frac{x}{d_-}$, which gives the desired result.

(e). From Lemma 8.1(e),

$$\lim_{t\to\infty} \frac{D(t)}{t} = \sum_{j\in E} d(j)\,\pi_j, \text{ a.s.},$$

and since, by Lemma 8.1(a) and (b), D has continuous and strictly increasing sample paths, then for all $x > 0$,

$$B(x) = t \Leftrightarrow D(t) = x.$$

Since, by (d), $\lim_{x\to\infty} B(x) = \infty$, this gives

$$\lim_{x\to\infty} \frac{B(x)}{x} = \lim_{t\to\infty} \frac{t}{D(t)} = \frac{1}{\sum_{j\in E} d(j)\,\pi_j}.$$

(f). The statement is a consequence of the results obtained in Example 8.3 in the next section. □

Remark 8.3. If we let J^B be the Markov chain with intensity matrix $\Sigma^B = D^{-1}\Lambda P$, where $D = (\delta_{jk} d(j))$, then (up to the explosion time $D(\infty)$ and by (8.39)) $(B, J \circ B)$ is equivalent to the MCPP (B, J^B) with drift $[d(j)]^{-1}$, $j \in E$, and no jumps. Note that if J is ergodic and $d_+ < \infty$ then J^B is also ergodic and has stationary distribution π^B given by:

$$\pi_j^B = \frac{\pi_j\, d(j)}{\sum_{k\in E} \pi_k\, d(k)};$$

this would lead to Theorem 8.4 (e) using a procedure similar to the one used to obtain Lemma 8.1 (e). Since the infinitesimal generator of a process characterizes the process, using Theorem 8.2, we obtain

$$B(x) = \int_0^x \frac{1}{d(J^B(y))}\, dy\,, \text{ for } x < D(\infty).$$

As was done for the demand, we can identify the inverse of the demand as a random clock with $[d(J^B(y))]^{-1}$ being its time speed at y.

8.5 The Busy-Period Process

The busy-period starting from level x, $x > 0$, is defined as the time the storage system would become empty if the initial level was x. Denoting this time as $T(x, k)$ when the initial state of J is k, we have

$$T(x,k) = \inf\,\{t : Z(t) = 0\} \quad \text{on} \quad \{Z(0) = x, J(0) = k\}, \tag{8.40}$$

for $k \in E$, with the convention that the infimum of an empty set is $+\infty$. We are also interested in the state of J at the time of emptiness, namely

$J \circ T(x,k)$. If $T(x,k) = \infty$ we define $J \circ T(x,k) = \Delta$. Of special interest is the situation when the initial level x has the distribution M_{jk} with $j \in E$. This busy period may represent (for example) the busy-period initiated by the input arriving with a transition from state j to state k in J. In this case, we shall call this the busy-period initiated by the transition j to k and denote it as T_{jk}. Let

$$G_{jk}(A,l) = P\{(T_{jk}, J \circ T_{jk}) \in A \times \{l\}\} \tag{8.41}$$

be the distribution of $(T_{jk}, J \circ T_{jk})$. We have

$$G_{jk}(A,l) = \int_{0^-}^{\infty} P\{(T, J \circ T)(x,k) \in A \times \{l\}\} \, M_{jk}\{dx\}, \tag{8.42}$$

for any Borel subset A of $[0,\infty], j,k \in E$ and $l \in E \cup \Delta$. Note that T_{jk} has possibly an atom at (∞, Δ). We let $G_{jkl}(A) = G_{jk}(A,l)$ and for simplicity we attach the atom at (∞, Δ) (with value $G_{jk\Delta}\{\infty\}$) to the measure G_{jkj}, so that, for $j,k,l \in E$ with j,k fixed,

$$\sum_{l \neq j} G_{jkl}(\mathbf{R}_+) + G_{jkj}([0,+\infty]) = 1. \tag{8.43}$$

We next consider the process $(T, J \circ T) = \{(T, J \circ T)(x),\ x \geq 0\}$ where $T(x)$ is now defined as $T(x, J(0))$ with $J(0)$ having an arbitrary distribution. Thus we are studying the effect of the initial storage level, regardless of the initial state of J. We have the following.

Lemma 8.3. *$(T, J \circ T)$ is a MAP with infinitesimal generator given by*

$$\begin{aligned} \mathcal{A}_{(T,J\circ T)}\, f(x,j) &= \frac{1}{d(j)} \frac{\partial}{\partial x} f(x,j) \\ &\quad + \frac{1}{d(j)} \sum_{l \in E} \int_{0^-}^{\infty^+} [f(x+y,l) - f(x,j)] \sum_{k \in E} \sigma_{jk} G_{jkl}\{dy\}, \end{aligned} \tag{8.44}$$

where $G_{jkl}\{\infty\} = 0$ for $l \neq j$.

Proof. On the set $\{J(0) = j\}$, $j \in E$, consider for $A \in \mathcal{B}([0,+\infty))$ the events

$$C(j,h;A,l) = \{T(h) \in A, J \circ T(h) = l\},$$

$$H(j,h;A,l) = \left\{ \begin{array}{l} T(h) \in A, J \circ T(h) = l \text{ and on } [0, B(h)] \\ \text{there are no transitions in } J \end{array} \right\},$$

and for $k \in E$,

$$E_k(j,h;A,l) = \left\{ \begin{array}{l} T(h) \in A, J \circ T(h) = l \text{ and the first transition in } J \\ \text{is to state } k \text{ and occurs before time } B(h) \end{array} \right\}.$$

It is easily seen that

$$P[C(j,h;A,l)] = \sum_{k\in E} P[E_k(j,h;A,l)] + P[H(j,h;A,l)]$$

$$= \sum_{k\in E} \int_0^{\frac{h}{d(j)}} \sigma_{jk} e^{-\lambda_j \tau} \int_{0-}^{\infty^+} \sum_{m\in E} G_{jkm}\{A-\tau-ds\}\, P[C(m,h-\tau d(j);ds,l)]d\tau$$

$$+ \delta_{jl}\left[1-\lambda_j \frac{h}{d(j)}\right] 1_{\{\frac{h}{d(j)}\in A\}} + o(h).$$

We are now able to obtain the infinitesimal generator of $(T, J\circ T)$,

$$\mathcal{A}_{(T,J\circ T)} f(x,j) = \lim_{h\to 0+}\left[\frac{1}{h}\sum_{l\in E}\int_{0-}^{\infty^+} [f(x+y,l)-f(x,j)]\, P[C(j,h;dy,l)]\right]$$

$$= \frac{1}{d(j)} \lim_{h\to 0+}\left[\int_{0-}^{\infty^+} \frac{f(x+y,j)-f(x,j)}{h/d(j)}\left(1-\lambda_j \frac{h}{d(j)}\right) \mathbf{1}_{\{\frac{h}{d(j)}\in dy\}}\right]$$

$$+ \sum_{l\in E}\int_{0-}^{\infty^+} [f(x+y,l)-f(x,j)] \sum_{k\in E}\frac{\sigma_{jk}}{d(j)}$$

$$\lim_{h\to 0+}\left[\frac{d(j)}{h}\int_0^{\frac{h}{d(j)}} e^{-\lambda_j\tau}\int_0^{y-\tau} \sum_{m\in E} G_{jkm}\{dy-\tau-ds\} P[C(m,h-\tau d(j);ds,l)]d\tau\right]$$

$$= \frac{1}{d(j)}\frac{\partial}{\partial x} f(x,j) + \sum_{l\in E}\int_{0-}^{\infty^+} [f(x+y,l)-f(x,j)] \sum_{k\in E}\frac{\sigma_{jk}}{d(j)} G_{jkl}\{dy\},$$

which is the desired result. □

Example 8.3. As an application of Lemma 8.3 we derive the infinitesimal generator of the inverse of the demand process $\{(B, J\circ B)(x),\ x\geq 0\}$ which coincides with the storage busy-period process if no input arrives to the system, i.e.

$$M_{jk} = \varepsilon_0, \quad \forall j,k\in E,$$

where ε_0 is a distribution concentrated at the origin. This implies that for $A\in\mathcal{B}([0,+\infty])$,

$$G_{jkl}(A) = \begin{cases} 0 & \text{if } k\neq l \\ \varepsilon_0(A) & \text{if } k=l \end{cases}, \tag{8.45}$$

so that the generator of $\{(B, J\circ B)(x),\ x\geq 0\}$ is given by

$$\mathcal{A}_{(B,J\circ B)} f(x,j) = \frac{1}{d(j)}\left[\frac{\partial f}{\partial x}(x,j) + \sum_{l\neq j}\sigma_{jl}[f(x,l)-f(x,j)]\right],$$

thus proving Theorem 8.4 (f).

We consider, for $j \in E$, the random variable $(\tilde{T}_j, J \circ \tilde{T}_j)$, where $\tilde{T}_j$ represents the duration of the storage busy-period initiated by a new input if at present J is in state j. Denote by $\tilde{G}_j$ its distribution and by $p^\star_{jl}$ the probability that this busy-period ends in state l. Our setting implies that for $A \in \mathcal{B}([0,\infty])$, and $l \in E$,

$$\tilde{G}_{jl}(A) \equiv \tilde{G}_j(A, \{l\}) = \sum_{k \in E} p_{jk} G_{jkl}(A), \tag{8.46}$$

and

$$p^\star_{jl} = P\{J \circ \tilde{T}_j = l\} = \tilde{G}_{jl}([0,\infty]). \tag{8.47}$$

In terms of the definitions above, the infinitesimal generator of $(T, J \circ T)$ (given by (8.44)) can be expressed, with $\tilde{G}_{jl}\{\infty\} = 0$ for $l \neq j$, as follows:

$$\begin{aligned} \mathcal{A}_{(T,J\circ T)}\, f(x,j) &= \frac{1}{d(j)} \frac{\partial f}{\partial x}(x,j) \\ &\quad + \frac{\lambda_j}{d(j)} \sum_{l \in E} \int_{0^-}^{\infty^+} [f(x+y,l) - f(x,j)]\, \tilde{G}_{jl}\{dy\}. \end{aligned} \tag{8.48}$$

We elaborate a bit more to increase the interpretability of (8.48). Suppose the present state of the Markov chain J is j. We are now able to consider the (conditional) distribution of the length of the busy-period initiated by a new input given that at the end of the same busy period the Markov chain J is in a given state l. We denote such a variable by $T^\star_{jl}$ and its distribution by $M^\star_{jl}$ and for completeness we set $T^\star_{jl} = 0$ if $p^\star_{jl} = 0$. Thus, for $A \in \mathcal{B}([0,\infty])$, and $j, l \in E$,

$$M^\star_{jl}(A) = \begin{cases} \frac{\tilde{G}_{jl}(A)}{\tilde{G}_{jl}([0,\infty])} & p^\star_{jl} \neq 0 \\ \mathbf{1}_{\{0 \in A\}} & p^\star_{jl} = 0 \end{cases}, \tag{8.49}$$

where we note that

$$p^\star_{jl} M^\star_{jl}(A) = \tilde{G}_{jl}(A). \tag{8.50}$$

We define for $j, l \in E$,

$$\lambda^\star_j = \frac{\lambda_j}{d(j)}, \qquad \sigma^\star_{jl} = \lambda^\star_j\, p^\star_{jl}, \tag{8.51}$$

and let $J^\star$ be a Markov chain on the state space E with intensity matrix $\Sigma^\star = D^{-1}\Lambda P^\star$, where $P^\star = (p^\star_{jk})$ and $D = (\delta_{jk}\, d(j))$. As a direct consequence of Lemma 8.3, or even more immediately from (8.48), we have the following result.

Theorem 8.5. *$(T, J \circ T)$ is a MAP whose generator is given by*

$$\mathcal{A}_{(T,J\circ T)} f(x,j) = \frac{1}{d(j)} \frac{\partial f}{\partial x}(x,j) + \int_{0-}^{\infty^+} [f(x+y,j) - f(x,j)]\, \sigma^\star_{jj} M^\star_{jj}\{dy\}$$
$$+ \sum_{l \neq j} \int_{0-}^{\infty^+} [f(x+y,l) - f(x,j)]\, \sigma^\star_{jl} M^\star_{jl}\{dy\}. \quad (8.52)$$

Thus $(T, J^\star)$ is an MCPP with drift function $[d(j)]^{-1}, j \in E$, and jumps with distribution $M^\star_{jl}$ occurring at rate $\sigma^\star_{jl}$.

Theorem 8.5 provides insights that lead to a more meaningful probabilistic interpretation of the results obtained in Lemma 8.3. Among other facts, it is easy to see that the motion of the process $(T, J^\star)$ is related with the motion of the process $(T, J \circ T)$, but the perspective from which we see the busy-period evolving is quite different in the two processes. Some comments are in order.

Remark 8.4. The transitions between states of E occurring during a busy-period initiated by a new input in $(T, J \circ T)$ are eliminated in the process $(T, J^\star)$. In addition, the jumps in the additive component of the process $(T, J^\star)$ are lengths of busy-periods initiated by new arrivals.

Remark 8.5. Time is rescaled; when $J^\star$ is in state j, time is measured in units $[d(j)]^{-1}$ if referred to the Markov chain J. Thus the busy-period process increases locally at rate equal to the reciprocal of the present demand rate $[d(j)]^{-1}$. In particular, the busy period local timing is only a function of the present state of the Markov chain $J^\star$.

Remark 8.6. With respect to the length of busy-periods initiated by new arrivals, the actual values of the jump rates $\{\lambda_j, j \in E\}$ and of the demand rates $\{d(j), j \in E\}$ are immaterial; these quantities impact the length of busy-periods initiated by new arrivals only through the ratios $\{\lambda^\star_j = \frac{\lambda_j}{d(j)}, j \in E\}$. This expresses a relativity property of the jump rates and demand rates. Nevertheless the demand rates impact the length of busy-periods (locally) as a function of their absolute values in terms of the drift of the busy period process.

As shown in Theorem 8.5, the storage busy-period process $(T, J^\star)$ is an MCPP, thus a relatively simple process. Nevertheless, at this point, we cannot make direct use of this property since we do not know the distributions $M^\star_{jl}$ of the jumps, nor the values of the probabilities $p^\star_{jl}$. In the following, our aim will be to get a better description of the objects we have just mentioned.

8.6 Some More Notations and Results

In this section we mainly introduce some additional notations and results that we use in the following sections.

Definition 8.2. If, for $j, l \in E$, $A_{jl}\{dx\}$ is a finite measure on $[0, \infty]$ and $\theta > 0$, then we denote by $A(\theta) = (A_{jl}(\theta))$ the Laplace transform matrix of the matrix measure $A = (A_{jl}\{dx\})$,

$$A(\theta) = \left(\int_0^\infty e^{-\theta x} A_{jl}\{dx\} \right). \tag{8.53}$$

For completeness, if $A = (A_{jl})$ is a matrix of nonnegative numbers we consider that A_{jl} is an atomic measure concentrated solely at the origin to which it assigns measure A_{jl}, so that $A(\theta) \equiv A$.

Definition 8.3. We let $S = \{f = (f_{jk}), j, k \in E$ such that $f_{jk} : \mathbf{R}_+ \to \mathbf{R}$ is bounded and continuous$\}$, and define a norm in S,

$$\|f\| = \sup_{k \in E} \sum_{j \in E} \|f_{jk}\|,$$

where $\|f_{jk}\| = \sup_{x \geq 0} \mid f_{jk}(x) \mid$. Then, $(S, \|.\|)$ is a Banach space. Moreover, we define, for $f \in S$ and $s > 0$,

$$\mathcal{L}(f, s) = \left(\int_0^\infty e^{-st} f_{jk}(t)\, dt \right).$$

Definition 8.4. Suppose J is a pure jump process on a countable state space E, that V and W are nonnegative processes such that (V, J) and (W, J) are adapted to a given filtration $\mathcal{F} = \{\mathcal{F}_t, t \geq 0\}$, and $T(x)$ is an $\mathcal{F}$-stopping time, for all $x \geq 0$. We define (with the operator E denoting the usual expectation and A being $\mathcal{F}_0$-measurable) the following types of expectation matrices:

$$E_J[W(t)] = (E[W(t); J(t) = k | J(0) = j]),$$

$$E_J[W(t)|A] = (E[W(t); J(t) = k | A,\, J(0) = j]),$$

$$E_J[T(x)] = (E[T(x); J \circ T(x) = k | J(0) = j]).$$

In a similar way, we define, for $\theta, \theta_1, \theta_2 > 0$, the following types of Laplace transform matrices:

$$\psi_W(s;\theta) = \mathcal{L}\left(E_J\left[e^{-\theta W(\cdot)}\right], s\right),$$

$$\psi_{(W,S)}(s;\theta_1,\theta_2) = \mathcal{L}\left(E_J\left[e^{-\theta_1 W(\cdot)-\theta_2 S(\cdot)}\right], s\right);$$

obvious definitions are considered for the conditional case (e.g. $\psi_{W|A}(s;\theta)$).

With $D = (d(j)\delta_{jk})$, we define for $\theta > 0$

$$\begin{cases} \Phi(\theta) = \Lambda\left[I - \Gamma(\theta)\right] \\ \Phi^\star(\theta) = D^{-1}\Lambda\left[I - \Gamma^\star(\theta)\right] \end{cases}, \tag{8.54}$$

where,

$$\begin{cases} \Gamma(\theta) = (p_{jk}M_{jk}(\theta)) \\ \Gamma^\star(\theta) = (p^\star_{jk}M^\star_{jk}(\theta)) \end{cases}, \tag{8.55}$$

and, in addition, we let

$$\begin{cases} \Psi(\theta) = \Phi(\theta) + D\theta \\ \Psi^\star(\theta) = \Phi^\star(\theta) + D^{-1}\theta \end{cases}. \tag{8.56}$$

Suppose J is a Markov chain on a countable state space E with intensity matrix Σ, and (W, J) is an MCPP with nonnegative jumps whose infinitesimal generator is given by

$$\mathcal{A}_{(W,J)}\, f(x,j) = d(j)\frac{\partial f}{\partial x}(x,j) + \sum_{k\in E}\sigma_{jk}\int_{0-}^{\infty}[f(x+y,k) - f(x,j)]M_{jk}\{dy\}. \tag{8.57}$$

The distribution measures M_{jk} are thus concentrated on $\mathbf{R}_+$. We assume that $\{d(j)\}$ and $\{\lambda_j = \sum_{k\in E}\sigma_{jk}\}$ are uniformly bounded.

Lemma 8.4. *The process W with generator given by Eq. (8.57) has the following Laplace transform matrix*

$$E_J\left[e^{-\theta W(t)}\right] = e^{-t\Psi_W(\theta)} = e^{-t[\Phi_W(\theta)+D_W\theta]}, \tag{8.58}$$

where $\Psi_W = \Psi$, $\Phi_W = \Phi$, $D_W = D$, so that

$$\Psi_W(\theta) = \Lambda\left[I - \Gamma(\theta)\right] + D\theta. \tag{8.59}$$

Proof. The result follows similarly to Theorem 2.16. □

Remark 8.7. The Laplace transform matrix of the additive component of an MCPP with nonnegative jumps is a direct extension (to a matrix-like form) of the Laplace transform of a compound Poisson process, as it is easily seen from equations (8.58) and (8.59).

The presence of the matrices Λ and D in equations (8.58) and (8.59) expresses the fact that the (compound) arrival rate and the drift of the additive component of an MCPP depend on the state of the underlying Markov chain, as opposed to the compound Poisson case where they are fixed constants.

Remark 8.8. Assume $d_+ < \infty$, $s, \theta > 0$, and let $Q = \Sigma - \Lambda$ and 0 denote the null process $0(t) \equiv 0$ then, using Lemma 8.4, we have:

$$\pi(t) = (P\left[J(t) = k | J(0) = j\right]) = E_J\left[e^{-\theta 0(t)}\right] = e^{t\,Q}, \tag{8.60}$$

$$\hat{\pi}(s) = \int_0^\infty e^{-st}\pi(t)\,dt = \psi_0(s;\cdot) = \left[sI - Q\right]^{-1}, \tag{8.61}$$

$$\begin{cases} E_J\left[e^{-\theta X(t)}\right] = e^{-t\Phi(\theta)} \\ E_J\left[e^{-\theta D(t)}\right] = e^{-t[D\theta - Q]} \\ E_J\left[e^{-\theta Y(t)}\right] = e^{-t[\Phi(\theta) - D\theta]} \end{cases}, \tag{8.62}$$

and

$$\begin{cases} \psi_X(s;\theta) = \left[sI + \Phi(\theta)\right]^{-1} \\ \psi_D(s;\theta) = \left[sI + D\theta - Q\right]^{-1} \\ \psi_Y(s;\theta) = \left[sI - D\theta + \Phi(\theta)\right]^{-1} \end{cases}. \tag{8.63}$$

The following is a result on the generator of a operator semigroup, which is Proposition 2.5 of Goldstein ([42], Chapter 1).

Lemma 8.5. *Let W be a Banach space and $B(W)$ be the space of all bounded linear operators from W to W. If $\eta \in B(W)$, then*

$$\alpha = \left\{\alpha(t) = e^{t\eta} \doteq \sum_{n=0}^{\infty} \frac{(t\eta)^n}{n!};\, t \in \mathbf{R}_+\right\} \tag{8.64}$$

is a semigroup satisfying:

$$\|\alpha(t) - I\| \to 0, \qquad \text{as } t \to 0^+. \tag{8.65}$$

Moreover, η is the generator of α. Conversely, if α is a semigroup satisfying (8.65), then the generator η of α belongs to $B(W)$ and $\alpha(t) = e^{t\eta}$.

Next we give a definition that is very useful for the remainder of the chapter.

Definition 8.5. If $-\eta$ is the generator of a semigroup as described in Lemma 8.5 and $A(\theta)$ is a Laplace transform matrix as described in Definition 8.2, we define

$$A \circ \eta = \left(\int_{0-}^{\infty} \sum_{k \in E} A_{jk}\{dx\} \left(e^{-x\eta}\right)_{kl} \right).$$

Remark 8.9. Note that if A is a matrix of nonnegative numbers, then $A \circ \eta = A$. Very important is the fact that, if in Definition 8.5 A is such that $(A_{jl}[0,\infty])$ is a sub-stochastic matrix and $-\eta$ is the generator of a contraction semigroup, then $A \circ \eta$ is a contraction map from S to S.

8.7 Laplace Transform of the Busy Period

The process $(T, J^{\star})$ is an MCPP with nonnegative jumps and positive drift. We now use the definitions and results of the previous section to derive the Laplace transform matrix of T, but in order to use Lemma 8.4 we need to add some additional conditions to the storage process, which we call the *standard conditions*,

$$0 < d_{-} \leq d_{+} < \infty \qquad \text{and} \qquad \lambda_{+} = \sup_{j \in E} \lambda_j < \infty. \tag{8.66}$$

In the rest of the chapter we assume implicitly that the *standard conditions* hold.

Theorem 8.6. *The Laplace transform matrix of the storage busy-period is given by*

$$E_J\left[e^{-\theta T(x)}\right] = e^{-x\Psi^{\star}(\theta)}, \tag{8.67}$$

where as previously defined,

$$\Psi^{\star}(\theta) = \Phi^{\star}(\theta) + D^{-1}\theta = D^{-1}\Lambda\left[I - \Gamma^{\star}(\theta)\right] + D^{-1}\theta. \tag{8.68}$$

Moreover, and more importantly, $\Gamma^{\star}$ *satisfies the functional equation*

$$\Gamma^{\star}(\theta) = \Gamma \circ \Psi^{\star}(\theta). \tag{8.69}$$

so that

$$\Gamma^{\star}(\theta) = \Gamma \circ \left(D^{-1}\theta + D^{-1}\Lambda\left[I - \Gamma^{\star}(\theta)\right]\right), \tag{8.70}$$

and

$$D\Psi^{\star}(\theta) = \theta I + \Phi \circ \Psi^{\star}(\theta). \tag{8.71}$$

Proof. From the conditions of the theorem, Theorem 8.5, and since (as a consequence of the standard conditions) the rates $\{\lambda_j^\star = \frac{\lambda_j}{d(j)}\}$ and $\{[d(j)]^{-1}\}$ are uniformly bounded, we conclude, by Lemma 8.4, that $E_J\left[e^{-\theta T(x)}\right] = e^{-x\Psi_T(\theta)}$, where

$$\Psi_T(\theta) = \Phi_T(\theta) + D_T\theta = D^{-1}\Lambda\left[I - \Gamma^\star(\theta)\right] + D^{-1}\theta = \Psi^\star(\theta)$$

and

$$\begin{aligned}\Gamma^\star(\theta) &= \left(\int_{0-}^{\infty}\sum_{k\in E} p_{jk}M_{jk}\{dx\}E\left[e^{-\theta T(x)}; J\circ T(x) = l|J(0) = k\right]\right)\\ &= \left(\int_{0-}^{\infty}\sum_{k\in E} p_{jk}M_{jk}\{dx\}\left(e^{-x\Psi^\star(\theta)}\right)_{kl}\right) = \Gamma\circ\Psi^\star(\theta).\end{aligned}$$

Therefore, equations (8.67), (8.68), and (8.69) follow, and (8.70) is an immediate consequence of equations (8.68) and (8.69). From (8.68), (8.69), and Remark 8.9, we have

$$D\Psi^\star(\theta) = \theta I + \Lambda\left[I - \Gamma\right]\circ\Psi^\star(\theta) = \theta I + \Phi\circ\Psi^\star(\theta).$$

□

As (8.68) shows, to study the distribution of the busy-period initiated by a given initial storage level all we need to study is the distribution of the busy-period initiated by new arrivals (which may or may not be associated with a transition on J). Such information is given in our study by $\Gamma^\star(\theta)$. We show in Theorem 8.7 how $\Gamma^\star(\theta)$ may be computed.

Theorem 8.7. *There exists a unique bounded continuous solution $\Gamma^\star(\theta)$ of the functional equation (8.70). Moreover, if for $A \in S$ we define*

$$\Psi^\star(A(\theta)) = D^{-1}\theta + D^{-1}\Lambda\left[I - A(\theta)\right], \tag{8.72}$$

and let $\Gamma_0^\star(\theta) = 0$ say, and

$$\Gamma_{n+1}^\star(\theta) = \Gamma\circ\Psi^\star(\Gamma_n^\star(\theta)), \quad n = 0, 1, 2, ...,$$

then

$$\Gamma^\star(\theta) = \lim_{n\to\infty}\Gamma_n^\star(\theta).$$

Proof. In view of (8.72), (8.70) can be written as

$$\Gamma^\star(\theta) = \Gamma\circ\Psi^\star(\Gamma^\star(\theta)),$$

which is a fixed point equation. Moreover, by Remark 8.9, $\Gamma\circ\Psi^\star$ is a contraction map from S to S. The rest of the proof now follows from a Banach fixed point theorem such as ([38], Theorem 3.8.2). □

Remark 8.10. Consider an $M/G/1$ system with arrival rate λ and let $G^\star$ and $B^\star$ be the Laplace-Stieltjes transforms of the service time and busy-period initiated by a new arrival, respectively. Let $\phi = \lambda(1 - G^\star)$ and $E[e^{-sT(x)}] = e^{-x\eta(s)}$, where $T(x)$ is the duration of a busy-period initiated by a given amount of work x. It should be noted that Eq. (8.70) is a generalization of Takács functional equation for the $M/G/1$ busy-period initiated by a new input,

$$B^\star(s) = G^\star(s + \lambda[1 - B^\star(s)]),$$

whereas Eq. (8.71) is a generalization of the functional equation associated with the busy-period initiated by a given amount of work,

$$\eta(s) = s + \phi\left(\eta(s)\right),$$

as given in [86].

Remark 8.11. From Theorem 8.6 and since $(B, J \circ B)$ is equal to $(T, J \circ T)$ if there is no input (i.e. the distribution measures M_{jk} are probability measures concentrated at the origin. In this case $\Phi(\theta) \equiv \Lambda - \Sigma = -Q$ and, by (8.68), $\Psi^\star(\theta) = D^{-1}(\theta I - Q))$, it follows that

$$\begin{cases} E_J\left[e^{-\theta B(x)}\right] = e^{-xD^{-1}[\theta I - Q]} \\ \psi_B(s;\theta) = \left[sI + D^{-1}\left(\theta I - Q\right)\right]^{-1} \end{cases}.$$

8.8 The Unsatisfied Demand and Demand Rejection Rate

We note that $I(t)$ is the local time at zero of the storage level, where the clock has variable speed; the clock speed being $d(j)$ when J is in state j. We let $\zeta(t)$ be the demand rejection rate at time t, so that

$$\zeta(t) = d(J(t))\,\mathbf{1}_{\{Z(t)=0\}}, \tag{8.73}$$

and

$$I(t) = \int_0^t \zeta(s)\, ds. \tag{8.74}$$

Theorem 8.8. *For $\theta, s > 0$, we have:*

$$\mathcal{L}\left(E_J\left[e^{-\theta I(\cdot)}\zeta(\cdot)|Z(0) = x\right], s\right) = e^{-x\Psi^\star(s)}\left[\theta I + \Psi^\star(s)\right]^{-1}. \tag{8.75}$$

Proof. We have, using the fact that $\zeta(t) = 0$ for $t < T(x)$,

$$\mathcal{L}\left(E_J\left[e^{-\theta I(\cdot)}\zeta(\cdot)|Z(0)=x\right], s\right)$$
$$= \left(E\int_{T(x)}^{\infty}\left[e^{-st-\theta I(t)}\zeta(t); J(t)=k|Z(0)=x, J(0)=j\right]dt\right)$$

Now we carry out the transformation $I(t) = \tau$. We have $t = T(x+\tau) = T(x) + T(\tau)\circ\phi_{T(x)}$ (where ϕ is a shift operator as in Definition 8.1) and $d\tau = dI(t) = \zeta(t)dt$. Therefore, the right hand side of the last expression becomes, with $t = T(x) + T(\tau)\circ\phi_{T(x)}$,

$$\left(E\int_0^{\infty}\left[e^{-sT(x)-s[T(\tau)\circ\phi_{T(x)}]-\theta\tau}; J(t)=k|Z(0)=x, J(0)=j\right]d\tau\right)$$
$$= \left(\sum_{l\in E}\left(E_J\left[e^{-sT(x)}\right]\right)_{jl}\left(\int_0^{\infty}e^{-\theta\tau}E_J\left[e^{-sT(\tau)}\right]d\tau\right)_{lk}\right)$$
$$= \left(\sum_{l\in E}\left(e^{-x\Psi^\star(s)}\right)_{jl}\left(\int_0^{\infty}e^{-\tau[\theta I+\Psi^\star(s)]}d\tau\right)_{lk}\right)$$
$$= \left(\sum_{l\in E}\left(e^{-x\Psi^\star(s)}\right)_{jl}\left([\theta I+\Psi^\star(s)]^{-1}\right)_{lk}\right)$$
$$= e^{-x\Psi^\star(s)}\left[\theta I+\Psi^\star(s)\right]^{-1}.$$

□

Corollary 8.1. *For $\theta, s > 0$, we have:*

$$\mathcal{L}\left(E_J\left[\zeta(\cdot)|Z(0)=x\right], s\right) = e^{-x\Psi^\star(s)}\left[\Psi^\star(s)\right]^{-1} \tag{8.76}$$

and

$$\mathcal{L}\left(E_J\left[\mathbf{1}_{\{Z(\cdot)=0\}}|Z(0)=x\right], s\right) = e^{-x\Psi^\star(s)}\left[\Psi^\star(s)\right]^{-1}D^{-1}. \tag{8.77}$$

Proof. By letting $\theta \to 0$ in Theorem 8.8, (8.76) follows. If $J(t) = k$, $\zeta(t) = d(k)\,\mathbf{1}_{\{Z(t)=0\}}$. Thus,

$$\mathcal{L}\left(E_J\left[\zeta(\cdot)|Z(0)=x\right], s\right) = \mathcal{L}\left(E_J\left[\mathbf{1}_{\{Z(\cdot)=0\}}|Z(0)=x\right]D, s\right)$$
$$= \mathcal{L}\left(E_J\left[\mathbf{1}_{\{Z(\cdot)=0\}}|Z(0)=x\right], s\right)D,$$

and (8.77) follows from this and (8.76).

□

8.9 The Storage Level and Unsatisfied Demand

In this section we study the process $((Z,I),J)$; we are thus interested in the storage level and unsatisfied demand processes along with their interaction. We start this study with a lemma.

Lemma 8.6. *We have, for $\theta_1, \theta_2 > 0$,*

$$e^{-\theta_1 Z(t)-\theta_2 I(t)} = e^{-\theta_1[Z(0)+Y(t)]} - (\theta_1+\theta_2)\int_0^t e^{-\theta_1[Y(t)-Y(\tau)]-\theta_2 I(\tau)}\zeta(\tau)\,d\tau. \quad (8.78)$$

Proof. For $\theta > 0$, we have,

$$\theta\int_0^t e^{-\theta I(\tau)}\zeta(\tau)\,d\tau = \left[-e^{-\theta I(\tau)}\right]_0^t = 1 - e^{-\theta I(t)}. \quad (8.79)$$

Since $Z(t) = Z(0) + Y(t) + I(t)$, using (8.79), we have for $\theta_1, \theta_2 > 0$,

$$\begin{aligned}
e^{-\theta_1 Z(t)-\theta_2 I(t)} &= e^{-\theta_1[Z(0)+Y(t)]}\, e^{-(\theta_1+\theta_2)I(t)} \\
&= e^{-\theta_1[Z(0)+Y(t)]}\left[1-(\theta_1+\theta_2)\int_0^t e^{-(\theta_1+\theta_2)I(\tau)}\zeta(\tau)\,d\tau\right] \\
&= e^{-\theta_1[Z(0)+Y(t)]} - (\theta_1+\theta_2)\int_0^t e^{-\theta_1[Y(t)-Y(\tau)]-\theta_2 I(\tau)}\zeta(\tau)\,d\tau.
\end{aligned}$$

□

Theorem 8.9. *We have, for $s, \theta_1, \theta_2 > 0$,*

$$\begin{aligned}
&\psi_{(Z,I)|Z(0)=x}(s;\theta_1,\theta_2) \\
&\quad = \{e^{-\theta_1 x} I - (\theta_1+\theta_2)\, e^{-x\Psi^\star(s)}\, [\theta_2 I + \Psi^\star(s)]^{-1}\}\, \psi_Y(s;\theta_1), \quad (8.80)
\end{aligned}$$

where, as given by (8.63), $\psi_Y(s;\theta_1) = [sI - D\theta_1 + \Phi(\theta_1)]^{-1}$.

Proof. Using Lemma 8.6 and since (Y,J) is independent of $Z(0)$, we conclude that

$$\begin{aligned}
&\psi_{(Z,I)|Z(0)=x}(s;\theta_1,\theta_2) = \psi_{x+Y}(s;\theta_1) \\
&-(\theta_1+\theta_2)\mathcal{L}\left(E_J\left[\int_0^{(\cdot)} e^{-\theta_1[Y(\cdot)-Y(\tau)]-\theta_2 I(\tau)}\zeta(\tau)\,d\tau | Z(0)=x\right], s\right). \quad (8.81)
\end{aligned}$$

Now, note that

$$\psi_{x+Y}(s;\theta_1) = e^{-\theta_1 x}\psi_Y(s;\theta_1), \quad (8.82)$$

and since $Y(t) - Y(\tau)$ and $(I, \zeta)(\tau)$ are independent, as are (Y, J) and $Z(0)$, by interchanging the order of integration in the expression defining the second term in (8.81), we get

$$\left[\mathcal{L}\left(E_J\left[\int_0^{(.)} e^{-\theta_1[Y(\cdot)-Y(\tau)]-\theta_2 I(\tau)}\zeta(\tau)\,d\tau|Z(0)=x\right],s\right)\right]_{jk}$$

$$= \sum_{l\in E}\left[\mathcal{L}\left(E_J\left[e^{-\theta_2 I(\cdot)}\zeta(\cdot)|Z(0)=x\right],s\right)\right]_{jl}$$

$$\int_\tau^\infty e^{-s(t-\tau)}\,E\left[e^{-\theta_1[Y(t)-Y(\tau)]};\,J(t)=k|J(\tau)=l\right]dt$$

$$= \sum_{l\in E}\left(e^{-x\Psi^\star(s)}\left[\theta_2 I + \Psi^\star(s)\right]^{-1}\right)_{jl}\left[\psi_Y(s;\theta_1)\right]_{lk}$$

$$= \left[e^{-x\Psi^\star(s)}\left[\theta_2 I + \Psi^\star(s)\right]^{-1}\psi_Y(s;\theta_1)\right]_{jk}, \tag{8.83}$$

by Theorem 8.8. The statement follows from (8.81), (8.82) and (8.83). □

Corollary 8.2. *For $s, \theta > 0$, we have:*

$$\psi_{Z|Z(0)=x}(s;\theta) = \{e^{-\theta x}I - \theta\,e^{-x\Psi^\star(s)}\left[\Psi^\star(s)\right]^{-1}\}\,\psi_Y(s;\theta), \tag{8.84}$$

and

$$\psi_{I|Z(0)=x}(s;\theta) = \left\{I - \theta\,e^{-x\Psi^\star(s)}\left[\theta I + \Psi^\star(s)\right]^{-1}\right\}\hat{\pi}(s). \tag{8.85}$$

Proof. (8.84) and (8.85) follow by letting $\theta_2 \to 0$ and $\theta_1 \to 0$, respectively, in Eq. (8.80). □

If at time 0 the storage system is empty, Theorem 8.9 and Corollary 8.2 characterize the process $(Y - m, -m, J)$, where m is the infimum of the net input. Corollary 8.3 gives this characterization.

Corollary 8.3. *For $s, \theta_1, \theta_2 > 0$, we have,*

$$\psi_{(Y-m,-m)}(s;\theta_1,\theta_2) = \{I - (\theta_1+\theta_2)\left[\theta_2 I + \Psi^\star(s)\right]^{-1}\}\,\psi_Y(s;\theta_1), \tag{8.86}$$

$$\psi_{Y-m}(s;\theta) = \{I - \theta\left[\Psi^\star(s)\right]^{-1}\}\,\psi_Y(s;\theta), \tag{8.87}$$

$$\psi_{-m}(s;\theta) = \{I - \theta\left[\theta I + \Psi^\star(s)\right]^{-1}\}\,\hat{\pi}(s). \tag{8.88}$$

The unsatisfied demand up to time t, $I(t)$, is zero if and only if the storage system is nonempty on $[0,t]$. This fact is used to deduce Theorem 8.10 which characterizes the distribution of $I(t)$ as t goes to infinity.

Theorem 8.10. *If J is ergodic and $x, y \geq 0$, then*

$$\left(\lim_{t\to\infty} P[I(t) \leq y, J(t) = k \mid Z(0) = x, J(0) = j]\right) = \left[I - e^{-(x+y)\Psi^\star(0)}\right] \Pi. \quad (8.89)$$

Proof. For $t, x, y \geq 0$, using Corollary 8.2 and a Tauberian theorem, we have:

$$\begin{aligned}
&\left(\lim_{t\to\infty} P[I(t) \leq y, J(t) = k \mid Z(0) = x, J(0) = j]\right) \\
&= \left(\lim_{t\to\infty} P(I(t) = 0, J(t) = k \mid Z(0) = x + y, J(0) = j)\right) \\
&= \lim_{s\to 0+} \left\{ s\, \psi_{\mathbf{1}_{\{I(\cdot)=0\}}|Z(0)=x+y}(s;\theta) \right\} \\
&= \lim_{s\to 0+} \left\{ s \lim_{\theta\to\infty} \psi_{I|Z(0)=x+y}(s;\theta) \right\} \\
&= \lim_{s\to 0+} \left[s \lim_{\theta\to\infty} \left\{ I - e^{-(x+y)\Psi^\star(s)} \left[I + \theta^{-1}\Psi^\star(s)\right]^{-1} \right\} \hat{\pi}(s) \right] \\
&= \left[I - e^{-(x+y)\Psi^\star(0)}\right] \lim_{s\to 0+} \{ s\, \hat{\pi}(s) \} \\
&= \left[I - e^{-(x+y)\Psi^\star(0)}\right] \Pi.
\end{aligned}$$

□

8.10 The Steady State

In this section we study the storage process in steady state when a *congestion measure* ρ is less than one. We assume the state space E of the Markov chain J is finite but some of our results may hold for countable E. Note that, since E is finite, the *standard conditions* hold. In addition, if $\tilde{J}$ is irreducible, we may assume without loss of generality that $\tilde{J}$ is aperiodic (by introducing self-transitions in J that bring no input to the storage system).

Let $\bar{d}$ be the *demand rate* and $\bar{\alpha}$ be the *offered load*,

$$\bar{d} = \sum_{k\in E} \pi_k d(k), \qquad \bar{\alpha} = \sum_{k\in E} \pi_k \alpha_k, \quad (8.90)$$

where α_k is the offered load when J is in state k,

$$\alpha_k = \lambda_k \sum_{l\in E} p_{kl}\, \beta_{kl}, \qquad \beta_{kl} = \int_0^\infty x\, M_{kl}\{dx\}, \quad (8.91)$$

β_{kl} being the average value of a jump associated with a transition from state k to state l in J, which we assume finite. We now define the *congestion measure* ρ as

$$\rho = \frac{\bar{\alpha}}{\bar{d}} \, . \tag{8.92}$$

Lemma 8.7. *If $\rho < 1$, E is finite, and $\tilde{J}$ is irreducible (and aperiodic without loss of generality), then:*

(a) $\frac{Y(t)}{t} \to \bar{\alpha} - \bar{d}$ (< 0) *a.s., for all initial distributions.*
(b) $\inf\{t > 0 : Z(t) = 0\} < \infty$ *a.s., for all initial distributions.*
(c) *For all $j \in E$, $\tilde{T}_j < \infty$ a.s.*
(d) $\int_0^t \mathbf{1}_{\{Z(s)=0\}}\, ds \to \infty$ *a.s. as $t \to \infty$, for all initial distributions.*
(e) *For all $j, k \in E$, $p_{jk} > 0 \Rightarrow p^\star_{jk} > 0$.*
(f) *The DTMC with transition probability matrix $P^\star = (p^\star_{jk})$ is persistent nonnull and we denote its limit distribution by $\pi^\star$.*
(g) *The Markov chain J° with intensity matrix $\Lambda P^\star$ is persistent nonnull and has limit distribution π°, with*

$$\pi^\circ_k = \frac{\frac{\pi^\star_k}{\lambda_k}}{\sum_{l \in E} \frac{\pi^\star_l}{\lambda_l}} . \tag{8.93}$$

(h) $(Z, J)(t) \xrightarrow{\mathcal{L}} (Z_\infty, J_\infty)$ *as $t \to \infty$, for all initial distributions, where J_∞ is the stationary version of J and $\xrightarrow{\mathcal{L}}$ denotes convergence in distribution.*
(i) $\frac{I(t)}{t} \to \bar{d} - \bar{\alpha}$ *a.s., for all initial distributions.*

Proof. (a). Since Y can be seen as an accumulated reward process of the ergodic Markov chain J the result follows immediately from a limit theorem for this type of processes.
(b). Suppose $Z(0) = x \geq 0$, and let $T_x = \inf\{t \geq 0 : Z(t) = 0\}$. We have,

$$Y(s) \leq -x \Longrightarrow T_x \leq s. \tag{8.94}$$

The statement follows from (a) and (8.94).
(c). The statement is an immediate consequence of (b).
(d). From (b) and (c), the storage level process returns an infinite number of times to zero a.s.; moreover, since each visit to the zero level has a duration which is stochastically greater or equal to an exponential random variable with mean $[\sup_{j \in E} \lambda_j]^{-1}$, the statement follows.
(e). If $p_{jk} > 0$, we have

$$p^\star_{jk} \geq p_{jk} \int_0^\infty e^{-\lambda_k \frac{x}{d(k)}}\, M_{jk}\{dx\} > 0.$$

(f). From (e) and the fact that $\tilde{J}$ is irreducible and aperiodic, it follows that the DTMC with transition probability matrix $P^{\star}$ is irreducible and aperiodic, and therefore persistent nonnull, since E is finite.
(g). The statement follows directly from (f) and the relation between the stationary distribution of an ergodic Markov chain and its associated DTMC.
(h). The statement follows from the relation between limit properties of our storage process and the imbedded Markov-Lindley process $Z_n = Z(T_{n+1}-)$, as given by van Doorn and Regterschot ([115], Theorem 1) and the limit properties of this Markov-Lindley process as given in Theorem 7.3.
(i). From (h), $\frac{Z(t)}{t} \to 0$, a.s. From this, (a), and the fact that $Z(t) = Z(0) + Y(t) + I(t)$, the statement follows. □

Remark 8.12. Since our aim is to study the steady state behavior of the storage process, and using Lemma 8.7 (b) and (h), we may without loss of generality assume $Z(0) = 0$ a.s. From Lemma 8.7 (d), J° can be identified as J on the time set $\mathcal{T}^{\circ}$,

$$\mathcal{T}^{\circ} = \{t \geq 0 : Z(t) = 0\}, \tag{8.95}$$

and we will assume J° has time set $\mathcal{T}^{\circ}$. J° gives thus the evolution of J when all busy-periods are removed from consideration.

Theorem 8.11. *If $\rho < 1$, E is finite and $\tilde{J}$ is irreducible (and aperiodic without loss of generality), then*

(a)

$$P(Z_\infty = 0) = (1-\rho)\frac{\sum_{k\in E} \pi_k\, d(k)}{\sum_{k\in E} \pi_k^{\circ}\, d(k)}. \tag{8.96}$$

(b) *For all initial distributions,*

$$\frac{I(t)}{t} \longrightarrow (1-\rho)\sum_{k\in E} \pi_k\, d(k) = (1-\rho)\,\bar{d}, \ a.s. \tag{8.97}$$

(c) *With ζ_∞ being the limit distribution of $\zeta(t)$ and $k \in E$, we have*

$$P\left(Z_\infty = 0, J_\infty = k\right) = P\left(Z_\infty = 0\right)\pi_k^{\circ}, \tag{8.98}$$

$$E\left[\zeta_\infty \mathbf{1}_{\{J_\infty = k\}}\right] = P\left(Z_\infty = 0\right)\pi_k^{\circ}\, d(k). \tag{8.99}$$

(d) *With, from (c), $E_J\left[\zeta_\infty\right] = P\left(Z_\infty = 0\right)\Pi^{\circ} D$, where $\Pi^{\circ} = \mathbf{e}\pi^{\circ}$, we have*

$$E_J\left[e^{-\theta Z_\infty}\right] = E_J\left[\zeta_\infty\right]\theta\left[D\theta - \Phi(\theta)\right]^{-1}. \tag{8.100}$$

Proof. (a). With $\mathcal{T}^\circ$ defined by (8.95) and $\mathcal{T}_t^\circ = [0,t] \bigcap \mathcal{T}^\circ$, we have, by Remark 8.12 and Lemma 8.7 (g),

$$\lim_{t\to\infty} \frac{I(t)}{\int_0^t \mathbf{1}_{\{Z(u)=0\}}\,du} = \lim_{t\to\infty} \frac{\int_{\mathcal{T}_t^\circ} d(J(u))\,du}{\int_{\mathcal{T}_t^\circ} 1\,du}$$
$$= \lim_{s\to\infty} \frac{\int_0^s d(J^\circ(u))\,du}{\int_0^s 1\,du} = \sum_{k\in E} \pi_k^\circ\, d(k), \text{ a.s.} \quad (8.101)$$

From Lemma 8.7 (h), and for all initial distributions,

$$\frac{1}{t}\int_0^t \mathbf{1}_{\{Z(u)=0\}}\,du \to P\left(Z_\infty = 0\right), \text{ a.s.} \quad (8.102)$$

Moreover, from Lemma 8.7 (i), (8.101), (8.102) and the fact that

$$\frac{I(t)}{t} = \frac{I(t)}{\int_0^t \mathbf{1}_{\{Z(u)=0\}}\,du} \times \frac{\int_0^t \mathbf{1}_{\{Z(u)=0\}}\,du}{t}, \quad (8.103)$$

we get

$$\bar{d} - \bar{\alpha} = P\left(Z_\infty = 0\right) \sum_{k\in E} \pi_k^\circ\, d(k),$$

from which the result follows.

(b). The statement is an immediate consequence of Lemma 8.7 (i).

(c). The equation (8.98) holds in view of the fact that

$$P\left(Z_\infty = 0, J_\infty = k\right) = P\left(Z_\infty = 0\right) P\left(J_\infty = k \mid Z_\infty = 0\right)$$
$$= P\left(Z_\infty = 0\right) \pi_k^\circ,$$

and (8.99) follows directly from (8.98).

(d). From (8.84) and by a Tauberian theorem, we have

$$E_J\left[e^{-\theta Z_\infty}\right] = \lim_{s\to 0+} s\,\psi_{Z|Z(0)=0}(s;\theta)$$
$$= \left\{\lim_{s\to 0+} s\left[\Psi^\star(s)\right]^{-1}\right\} \theta\left[D\theta - \Phi(\theta)\right]^{-1}$$
$$= E_J\left[\zeta_\infty\right] \theta\left[D\theta - \Phi(\theta)\right]^{-1},$$

where the last inequality follows from (8.76) and a Tauberian theorem. □

Remark 8.13. Note that (8.100) is a generalization of the Pollaczek-Khinchin transform formula for the steady-state waiting time in queue in the $M/G/1$ system that, with the notation of Remark 8.10, $\rho = \lambda/\mu$ $(1/\mu$

being the expected value of the service time distribution), and $W^{\star}$ being the Laplace transform of the waiting time in queue, is given by:

$$W^{\star}(\theta) = (1-\rho)\,\frac{\theta}{\theta - \lambda\,[1 - G^{\star}(\theta)]}\,. \tag{8.104}$$

In particular, (8.100) and (8.104) show that in the Pollaczek-Khinchin formula $(1-\rho)$ stands for the limit demand rejection rate. Moreover, from (8.99), (8.96) and (8.97), the average demand rejection rate is $(1-\rho)\,\bar{d}$ which is (obviously) the limit rate of increase of $I(t)$. As (8.96) shows the limit probability of the storage system being empty may differ substantially from $(1-\rho)$.

Chapter 9

A Markovian Storage Model

In this chapter we investigate a storage model where the input and the demand are additive functionals on a Markov chain J. The storage policy is to meet the largest possible portion of the demand. We first derive results for the net input process imbedded at the epochs of transitions of J, which is a Markov random walk. Our analysis is based on a Wiener-Hopf factorization for this random walk; this also gives results for the busy-period of the storage process. The properties of the storage level and the unsatisfied demand are then derived.

9.1 Introduction

In this chapter we investigate the storage model in which the storage level $Z(t)$ at time t satisfies a.s. the integral equation

$$Z(t) = Z(0) + \int_0^t a(J(s))\,ds - \int_0^t r(Z(s), J(s))\,ds \tag{9.1}$$

where

$$r(x,j) = \begin{cases} d(j) & \text{if } x > 0 \\ \min(a(j), d(j)) & \text{if } x \le 0 \end{cases} \tag{9.2}$$

with the condition $Z(0) \ge 0$. Here $J = \{J(t),\, t \ge 0\}$ is a non-explosive Markov chain on a countable state space E, and a and d are nonnegative functions on E. The Eq. (9.1) states that when the Markov chain J is in state j at time t, input into the storage (buffer) occurs at rate $a(j)$, while

[9]This chapter is based on the article by A. Pacheco and N.U. Prabhu, 'A Markovian storage model', *Ann. Appl. Probab.* **6**, 1, pp. 76–91, 1996, and contents are reproduced with permission of the publisher, the Applied Probability Trust.

the demand occurs at rate $d(j)$ and the storage policy is to meet the largest possible portion of this demand. Let us denote by

$$X(t) = \int_0^t a(J(s))\,ds \tag{9.3}$$

and

$$D(t) = \int_0^t d(J(s))\,ds \tag{9.4}$$

the input and the (actual) demand during a time interval $(0, t]$. It can be seen that $(X, J) = \{(X, J)(t), t \geq 0\}$ and $(D, J) = \{(D, J)(t), t \geq 0\}$ are Markov-additive processes (MAPs) on the state space $\mathbf{R}_+ \times E$.

A storage model with a more general input process (X, J) of the Markov-additive type has been investigated in the previous chapter, where $X(t)$ consists of jumps of positive size as well as a (cumulative) drift $X(t)$. However, the analysis of the previous chapter cannot be applied in the present model, as there the assumption is made that $a(j) < d(j)$, for $j \in E$. This assumption would make (9.1) trivial since then the storage level would be decreasing, so that it would eventually reach zero after a random length of time and remain at zero after that. Therefore, in this chapter we do not assume that $a(j) < d(j)$, $j \in E$, and use a completely different approach to analyze our model. For this, we denote

$$\begin{cases} E_0 = \{j \in E : a(j) > d(j)\} \\ E_1 = \{j \in E : a(j) \leq d(j)\} \end{cases}. \tag{9.5}$$

The model represented by (9.1) occurs in data communication systems. Virtamo and Norros [116] have investigated a model in which a buffer receives input of data from a classical $M/M/1$ queueing system at a constant rate c_0 so long as the system is busy, and transmits these data at a maximum rate c_1 $(< c_0)$. Denoting by $J(t)$ the queue length, we can represent the buffer content $Z(t)$ at time t by

$$Z(t) = Z(0) + \int_0^t c_0 \mathbf{1}_{\{J(s)>0\}}\,ds - \int_0^t c_1 \mathbf{1}_{\{[J(s)>0]\vee[Z(s)>0]\}}\,ds. \tag{9.6}$$

This equation is of type (9.1) with

$$a(j) = \begin{cases} c_0 & \text{if } j > 0 \\ 0 & \text{if } j = 0 \end{cases} \qquad \text{and} \qquad d(j) = c_1,\ j \geq 0.$$

Clearly $a(0) < d(0)$, but $a(j) > d(j)$ for $j > 0$. Thus $E_0 = \{1, 2, \ldots\}$ and $E_1 = \{0\}$. The input during $(0, t]$ is given by $c_0 B(t)$, where

$$B(t) = \int_0^t \mathbf{1}_{\{J(s)>0\}}\,ds,$$

this being the part of the time interval $(0, t]$ during which the server is busy. Other storage models of data communication systems satisfying Eq. (9.1) were presented in the previous chapter. A brief description of two of these models is the following.

Anick, Mitra and Sondhi [2] study a model for a data-handling system with N sources and a single transmission channel. The input rate is $a(j) = j$, where j is the number of sources that are "on", and the maximum output rate is a constant c, so that $d(j) = c$. We see that in this case $E_0 = \{\lfloor c \rfloor + 1, \lfloor c \rfloor + 2, \ldots\}$ and $E_1 = \{0, 1, \ldots, \lfloor c \rfloor\}$, with $\lfloor x \rfloor$ denoting the integer part of x.

Gaver and Lehoczky [40] investigate a model for an integrated circuit and packet switching multiplexer, with input of data and voice calls. There are $s + u$ output channels, of which s are for data transmission, while the remaining u are shared by data and voice calls (with calls having preemptive priority over data). Here calls arrive in a Poisson process and have exponentially distributed holding times. The model gives rise to Eq. (9.1) for the data buffer content $Z(t)$ with

$$\begin{cases} a(j) = c_0 \\ d(j) = c_2(s + j) \end{cases}$$

for $j = 0, 1, \ldots, u$, where j is the number of channels out of u not occupied by calls, c_0 is the (constant) data arrival rate, and c_2 is the output rate capacity per channel. Here

$$E_1 = \left\{ j \in \mathbf{N}_+ : \frac{c_0}{c_2} - s \leq j \leq u \right\} \quad \text{and} \quad E_0 = \left\{ j \in \mathbf{N}_+ : j < \frac{c_0}{c_2} - s \right\}.$$

The net input of our model $Y(t)$ is given by

$$Y(t) = A(t) - D(t) = \int_0^t y(J(s))\, ds \tag{9.7}$$

with $y = a - d$. The net input process is thus an MAP which is nonincreasing during periods in which the environment J is in E_1 and is increasing when J is in E_0. Its sample functions are continuous a.s., and differentiable everywhere except at the transition epochs $(T_n)_{n \geq 0}$ of the Markov chain J. Since

$$\int_0^t r(Z(s), J(s))\, ds = \int_0^t d(J(s))\, ds + \int_0^t \min\{y(J(s)), 0\}\, \mathbf{1}_{\{Z(s) \leq 0\}}\, ds,$$

we can rewrite Eq. (9.1) in the form

$$Z(t) = Z(0) + Y(t) + \int_0^t y(J(s))^- \, \mathbf{1}_{\{Z(s) \leq 0\}}\, ds \tag{9.8}$$

where $y^- = \max\{-y, 0\}$. Here the integral

$$I(t) = \int_0^t y(J(s))^- \mathbf{1}_{\{Z(s)\leq 0\}}\, ds \tag{9.9}$$

represents the amount of unsatisfied demand during $(0, t]$. If we let $J_n = J(T_n)$, then from Eq. (9.8) we obtain, for $T_n \leq t \leq T_{n+1}$,

$$Z(t) = Z(T_n) + y(J_n)(t - T_n) + y(J_n)^- \int_{T_n}^t \mathbf{1}_{\{Z(s)\leq 0\}}\, ds \tag{9.10}$$

and from (9.9)

$$I(t) = I(T_n) + y(J_n)^- \int_{T_n}^t \mathbf{1}_{\{Z(s)\leq 0\}}\, ds. \tag{9.11}$$

This shows that in order to study the process (Z, I, J) it may be of interest to first study the properties of the imbedded process $(Z(T_n), I(T_n), J(T_n))$.

The following is a brief summary of the results of this chapter. In Sec. 9.2 we give the solution of the integral Eq. (9.8); a particular consequence of the solution is that $Z(T_n)$ and $I(T_n)$ may be identified as functionals on the process $(T_n, Y(T_n), J(T_n))$, which is a Markov random walk (MRW). So the properties of this MRW are investigated in Sec. 9.3, the key result being a Wiener-Hopf factorization due to [97]; see Sec. 4.8. The results of Sec. 9.3 are used in Sec. 9.4 to study the properties of the storage level and the unsatisfied demand.

Rogers [99] investigates the model considered in the chapter, with the Markov chain having finite state space. His analysis is based on the Wiener-Hopf factorization of finite Markov chains, from which the invariant distribution of the storage level is derived. Methods for computing the invariant law of the storage level are discussed by Rogers and Shi [100]. Asmussen [11] and Karandikar and Kulkarni [49] investigate a storage process identified as the reflected Brownian Motion (BM) modulated by a finite state Markov chain. In the case where the variance components of this BM are all zero their storage process reduces to the one considered in the chapter.

Our analysis allows for the Markov chain to have infinite state space and we derive the time dependent as well as the steady state behavior of both the storage level and the unsatisfied demand (see the two examples of the chapter). It should be noted that the specificity of our net input (namely, piecewise linearity) is not too relevant for our analysis, so the techniques used in the chapter can be applied to other net inputs.

We shall denote by $N = (\nu_{jk})$ the generator matrix of J and assume that J has a stationary distribution $(\pi_j,\ j \in E)$. For analytical convenience we assume that $y(j) \neq 0$ for $j \in E$.

9.2 Preliminary Results

We start by solving the integral Eq. (9.8). Proceeding as in the proof of Theorem 8.1, we have the following result.

Lemma 9.1. *We are given a stochastic process J, as defined above, on a probability space $(\Omega, \mathcal{F}, \mathcal{P})$, and additive functionals X and D on J as given in (9.3)-(9.4). The integral equation Eq. (9.8) with $Z(0) \geq 0$ a.s., has $\mathcal{P}$-a.s. the unique solution*

$$Z(t) = Z(0) + Y(t) + I(t), \tag{9.12}$$

where

$$I(t) = [Z(0) + m(t)]^- \tag{9.13}$$

with

$$m(t) = \inf_{0 \leq \tau \leq t} Y(\tau). \tag{9.14}$$

One of the consequences of the solution (9.12) is (9.10) since

$$Z(t) = \max\{Z(0) + Y(t), Y(t) - m(t)\} \geq [Z(0) + Y(t)]^+ \geq 0, \tag{9.15}$$

where $y^+ = \max\{y, 0\}$. We next prove some preliminary results concerning the imbedded process $(Z(T_n), I(T_n), J(T_n))$. We let

$$\begin{cases} Z_n = Z(T_n) \\ I_n = I(T_n) \\ S_n = Y(T_n) \\ Y_{n+1} = S_{n+1} - S_n \end{cases} \tag{9.16}$$

so that $Y_{n+1} = y(J_n)(T_{n+1} - T_n)$.

Lemma 9.2. *For $T_n \leq t \leq T_{n+1}$ we have*

$$\begin{cases} Z(t) = [Z_n + y(J_n)(t - T_n)]^+ \\ I(t) = I_n + [Z_n + y(J_n)(t - T_n)]^- \end{cases}, \tag{9.17}$$

where

$$\begin{cases} Z_n = Z_0 + S_n + I_n \\ I_n = [Z_0 + m_n]^- \end{cases} \tag{9.18}$$

with

$$m_n = \min_{0 \leq r \leq n} S_r. \tag{9.19}$$

Proof. Using (9.10) we may conclude that

$$Z(t) = \max\{0, Z_n + y(J_n)(t - T_n)\} \qquad (n \geq 0,\ t \in [T_n, T_{n+1}]),$$

which proves (9.17). As a consequence $Z_{n+1} = \max\{0, Z_n + Y_{n+1}\}$ for $n \geq 0$, which implies (9.18) in view of a result familiar in queueing systems (see, e.g., ([86], Chap. 2, Theorem 8)). □

Lemma 9.2 shows that in order to study the process (Z_n, I_n, J_n) it suffices to investigate the MRW (T_n, S_n, J_n), which we do in the next section. We note that if $T_n \leq t \leq T_{n+1}$, then $Y(t) = S_n + y(J_n)(t - T_n)$. Thus $\min(S_n, S_{n+1}) \leq Y(t) \leq \max(S_n, S_{n+1})$, which in turns implies that a.s. $\liminf Y(t) = \liminf S_n$ and $\limsup Y(t) = \limsup S_n$. Similarly, if we denote for $t \geq 0$ and $n = 0, 1, \ldots,$

$$M(t) = \sup_{0 \leq \tau \leq t} Y(\tau) \quad \text{and} \quad M_n = \max_{0 \leq r \leq n} S_r, \tag{9.20}$$

we may conclude that

$$\lim_{t \to \infty} M(t) = \lim_{n \to \infty} M_n = M \leq +\infty \tag{9.21}$$

$$\lim_{t \to \infty} m(t) = \lim_{n \to \infty} m_n = m \geq -\infty. \tag{9.22}$$

These statements show that some conclusions about the fluctuation behavior of the net input process may be drawn from the associated MRW (S_n, J_n). This in turn has implications for the storage level and unsatisfied demand since these processes depend on the net input. We denote by $(\pi_j^\star,\ j \in E)$ the stationary distribution of (J_n), so that

$$\pi_j^\star = \frac{(-\nu_{jj})\,\pi_j}{\sum_{k \in E} (-\nu_{kk})\,\pi_k}. \tag{9.23}$$

Also, define the *net input rate* $\overline{y} = \sum_{j \in E} \pi_j\, y(j)$, where we assume the sum exists, but may be infinite. We have then the following.

Theorem 9.1. (Fluctuation behavior of $Y(t)$). *We have a.s.*

(a) $\frac{Y(t)}{t} \to \overline{y}$.

(b) *If* $\overline{y} > 0$, *then* $\lim Y(t) = +\infty$, $m > -\infty$, *and* $M = +\infty$.

(c) *If* $\overline{y} = 0$, *then* $\liminf Y(t) = -\infty$, $\limsup Y(t) = +\infty$, $m = -\infty$, *and* $M = +\infty$.

(d) *If* $\overline{y} < 0$, *then* $\lim Y(t) = -\infty$, $m = -\infty$, *and* $M < +\infty$.

Proof. The proof of (a) is standard, but is given here for completeness. We have

$$\frac{Y(t)}{t} = \frac{1}{t}\int_0^t y(J(s))\,ds = \frac{1}{t}\int_0^t \{[y(J(s))]^+ - [y(J(s))]^-\}\,ds. \tag{9.24}$$

Now since J is ergodic

$$\begin{aligned}\frac{1}{t}\int_0^t [y(J(s))]^+\,ds &= \sum_{j\in E}\frac{[y(j)]^+}{t}\int_0^t \mathbf{1}_{\{J(s)=j\}}\,ds \\ &\to \sum_{j\in E}[y(j)]^+\,\pi_j\ ,\ \text{as } t\to\infty,\ \text{a.s.}\end{aligned} \tag{9.25}$$

and similarly $\frac{1}{t}\int_0^t [y(J(s))]^-\,ds \to \sum_{j\in E}[y(j)]^-\,\pi_j$ a.s. as $t\to\infty$, so that using (9.24) we conclude that $\lim_{t\to\infty} Y(t)/t = \overline{y}$ a.s. Define the mean increment in the MRW (S_n, J_n)

$$\mu^\star = \sum_{j\in E}\pi_j^\star\, E\left[Y_1 \mid J_0 = j\right]. \tag{9.26}$$

Since $E\left[Y_1 \mid J_0 = j\right] = y(j)/(-\nu_{jj})$ it follows that $\overline{y} = \mu^\star \sum_{k\in E}(-\nu_{kk})\,\pi_k$. The statements (b)-(d) follow from this and (9.21)-(9.22), by using Theorem 4.8 and Proposition 7.2. These two last results describe the fluctuation behavior of the MRW (S_n, J_n). □

9.3 The Imbedded MRW

In this section we investigate the properties of the MRW (T_n, S_n, J_n). We note that the conditional distribution of the increments $(T_n - T_{n-1}, S_n - S_{n-1})$ given J_{n-1} is singular, since $Y_n = S_n - S_{n-1} = y(J_n)(T_n - T_{n-1})$ a.s. The distribution of (T_1, Y_1, J_1) is best described by the transform matrix

$$\Phi(\theta,\omega) = (\phi_{jk}(\theta,\omega)) = \left(E\left[e^{-\theta T_1 + i\omega Y_1}; J_1 = k \mid J_0 = j\right]\right) \tag{9.27}$$

for $\theta > 0$, ω real and $i = \sqrt{-1}$. We find that

$$\begin{aligned}\Phi(\theta,\omega) &= (\phi_{jk}(\theta,\omega)) = (\alpha_j(\theta,\omega)\,p_{jk}) \\ &= (\alpha_j(\theta,\omega)\,\delta_{jk})\,(p_{jk}) = \alpha(\theta,\omega)P\end{aligned} \tag{9.28}$$

where

$$\alpha_j(\theta,\omega) = \frac{-\nu_{jj}}{-\nu_{jj} + \theta - i\omega y(j)} \quad \text{and} \quad p_{jk} = \begin{cases} -\frac{\nu_{jk}}{\nu_{jj}} & k\neq j \\ 0 & j = k\end{cases}. \tag{9.29}$$

For the time-reversed MRW $(\widehat{T}_n, \widehat{S}_n, \widehat{J}_n)$ corresponding to the given MRW we have

$$\widehat{\Phi}(\theta,\omega) = \left(\widehat{\phi}_{jk}(\theta,\omega)\right) = \left(E\left[e^{-\theta\widehat{T}_1+i\omega\widehat{Y}_1}; \widehat{J}_1 = k \mid \widehat{J}_0 = j\right]\right)$$
$$= \left(\frac{\pi_k^\star}{\pi_j^\star} E\left[e^{-\theta T_1+i\omega Y_1}; J_1 = j \mid J_0 = k\right]\right) = \widehat{P}\alpha(\theta,\omega) \quad (9.30)$$

where $\widehat{P}$ is the transition probability matrix of the time-reversed chain $\widehat{J}$, namely

$$\widehat{P} = (\widehat{p}_{jk}) = \left(\frac{\pi_k^\star}{\pi_j^\star} p_{kj}\right). \quad (9.31)$$

Since the T_n are non-decreasing a.s., the fluctuating theory of the MRW (T_n, S_n, J_n) is adequately described by (S_n). We now define the descending ladder epoch $\overline{N}$ of this MRW (T_n, S_n, J_n) and the ascending ladder epoch N of the time-reversed MRW $(\widehat{T}_n, \widehat{S}_n, \widehat{J}_n)$,

$$\begin{cases} \overline{N} = \min\{n : S_n < 0\} \\ N = \min\{n : \widehat{S}_n > 0\} \end{cases} . \quad (9.32)$$

(Here we adopt the convention that the minimum of an empty set is $+\infty$.) It should be noted that both N and $\overline{N}$ are strong ladder epochs, which is reasonable since the increments of S_n and $\widehat{S}_n$ in each case have an absolutely continuous distribution. The random variables $S_{\overline{N}}$ and $\widehat{S}_N$ are the ladder heights corresponding to $\overline{N}$ and N. We also denote the transforms (in matrix form)

$$\overline{\chi} = (\overline{\chi}_{jk}(z,\theta,\omega)) = \left(E\left[z^{\overline{N}} e^{-\theta T_{\overline{N}}+i\omega S_{\overline{N}}}; J_{\overline{N}} = k \mid J_0 = j\right]\right) \quad (9.33)$$

$$\chi = (\chi_{jk}(z,\theta,\omega)) = \left(\frac{\pi_k^\star}{\pi_j^\star} E\left[z^N e^{-\theta\widehat{T}_N+i\omega\widehat{S}_N}; \widehat{J}_N = j \mid \widehat{J}_0 = k\right]\right) \quad (9.34)$$

where $0 < z < 1$, $\theta > 0$, $i = \sqrt{-1}$, and ω is real. Connecting these two transforms is the Wiener-Hopf factorization, first established by Presman [97] analytically, and interpreted in terms of the ladder variables defined above by Prabhu, Tang and Zhu [95], as described in Chap. 4. The result is the following.

Lemma 9.3 (Wiener-Hopf factorization). *For the MRW (T_n, S_n, J_n) with $0 < z < 1$, $\theta > 0$, and ω real,*

$$I - z\Phi(\theta,\omega) = [I - \chi(z,\theta,\omega)]\,[I - \overline{\chi}(z,\theta,\omega)]. \quad (9.35)$$

We shall use this factorization and the special structure of our MRW to indicate how the transforms χ and $\overline{\chi}$ can be computed in the general case. It turns out that our results contain information concerning the descending ladder epoch $\overline{T}$ of the net input process (Y, J) and the ascending ladder epoch T of the time-reversed process $(\widehat{Y}, \widehat{J})$, which is defined as follows:

$$\widehat{J}(t) = \widehat{J}_n,\ \widehat{T}_{n-1} < t \leq \widehat{T}_n \quad \text{and} \quad \widehat{Y}(t) = \int_0^t y(\widehat{J}(s))\,ds. \tag{9.36}$$

Thus,

$$\begin{cases} \overline{T} = \inf\{t > 0 \,:\, Y(t) \leq 0\} \\ T = \inf\{t > 0 \,:\, \widehat{Y}(t) \geq 0\} \end{cases}. \tag{9.37}$$

We note that $Y(\overline{T}) = 0$ and $\widehat{Y}(T) = 0$ a.s. For $0 < z < 1$ and $\theta > 0$, we define the transforms

$$\zeta = (\zeta_{jk}(z, \theta)) = \left(E\left[z^{\overline{N}} e^{-\theta \overline{T}}; J(\overline{T}) = k \mid J(0) = j\right]\right) \tag{9.38}$$

$$\eta = (\eta_{jk}(z, \theta)) = \left(\frac{\pi_k^\star}{\pi_j^\star} E\left[z^N e^{-\theta T}; \widehat{J}(T) = j \mid \widehat{J}(0) = k\right]\right). \tag{9.39}$$

Theorem 9.2. *For $0 < z < 1$, $\theta > 0$, and ω real, we have*

$$\begin{cases} \overline{\chi}(z, \theta, \omega) = \zeta(z, \theta)\, \Phi(\theta, \omega) \\ \chi(z, \theta, \omega) = \alpha(\theta, \omega)\, \eta(z, \theta) \end{cases}. \tag{9.40}$$

Proof. An inspection of the sample paths of (Y, J) will show that $J(\overline{T}) = J_{\overline{N}-1}$ and

$$T_{\overline{N}} - \overline{T} = \frac{S_{\overline{N}}}{y(J(\overline{T}))} \quad \text{a.s.}$$

Since $\overline{T}$ is a stopping time for (Y, J), we see that given $J(\overline{T}) = l$, $S_{\overline{N}}/y(J(\overline{T}))$ is independent of $\overline{T}$ and has the same distribution as T_1, given $J_0 = l$. Therefore

$$\begin{aligned}
\overline{\chi}_{jk}(z, \theta, \omega) &= \sum_{l \in E} E[z^{\overline{N}} \exp(-\theta[\overline{T} + S_{\overline{N}}/y(J_{\overline{N}-1})] + i\omega S_{\overline{N}}); \\
&\qquad\qquad\qquad\qquad J_{\overline{N}-1} = l, J_{\overline{N}} = k \mid J_0 = j] \\
&= \sum_{l \in E} E[z^{\overline{N}} e^{-\theta \overline{T}}; J(\overline{T}) = l \mid J_0 = j] \\
&\qquad \cdot E\left[\exp\left(-\theta S_{\overline{N}}/y(J_{\overline{N}-1}) + i\omega S_{\overline{N}}\right); J_{\overline{N}} = k \mid J_{\overline{N}-1} = l\right] \\
&= \sum_{l \in E} \zeta_{jl}(z, \theta) E\left[e^{-\theta T_1 + i\omega Y_1}; J_1 = k \mid J_0 = l\right] \\
&= \sum_{l \in E} \zeta_{jl}(z, \theta)\, \phi_{lk}(\theta, \omega).
\end{aligned}$$

Thus $\overline{\chi}(z,\theta,\omega)=\zeta(z,\theta)\,\Phi(\theta,\omega)$. The proof of $\chi(z,\theta,\omega)=\alpha(\theta,\omega)\,\eta(z,\theta)$ is similar. □

In general, for an $(|E|\times|E|)$-matrix A we block-partition A in the form

$$A=\begin{pmatrix} A_{00} & A_{01} \\ A_{10} & A_{11}\end{pmatrix}$$

with the rows and columns of A_{00} corresponding to the states in E_0. We have now

$$\zeta=\begin{pmatrix} \mathbf{0} & \zeta_{01} \\ \mathbf{0} & \zeta_{11}\end{pmatrix},\qquad \eta=\begin{pmatrix} \eta_{00} & \eta_{01} \\ \mathbf{0} & \mathbf{0}\end{pmatrix},\qquad I=\begin{pmatrix} I_{00} & \mathbf{0} \\ \mathbf{0} & I_{11}\end{pmatrix}, \tag{9.41}$$

where I is the identity matrix. From (9.35) and Theorem 9.2 we have the following.

Theorem 9.3. *We have, for* $0<z<1$, $\theta>0$, *and* ω *real,*

$$\chi=\begin{pmatrix} \alpha_{00}\eta_{00} & \alpha_{00}\eta_{01} \\ \mathbf{0} & \mathbf{0}\end{pmatrix},\qquad \overline{\chi}=\begin{pmatrix} \zeta_{01}\Phi_{10} & \zeta_{01}\Phi_{11} \\ z\Phi_{10} & z\Phi_{11}\end{pmatrix}, \tag{9.42}$$

and

$$I_{00}-\overline{\chi}_{00}=(I_{00}-\chi_{00})^{-1}\left[(I_{00}-z\Phi_{00})-z\chi_{01}\Phi_{10}\right] \tag{9.43}$$

$$\overline{\chi}_{01}=(I_{00}-\chi_{00})^{-1}\left[z\Phi_{01}-\chi_{01}(I_{11}-z\Phi_{11})\right] \tag{9.44}$$

where the inverse exists in the specified domain.

We note that if we let

$$\overline{\gamma}=(\overline{\gamma}_{jk}(z,\theta))=\left(E\left[z^{\overline{N}}e^{-\theta T_{\overline{N}}};J_{\overline{N}-1}=k\mid J_0=j\right]\right) \tag{9.45}$$

$$\gamma=(\gamma_{jk}(z,\theta))=\left(\frac{\pi_k^\star}{\pi_j^\star}E\left[z^{N}e^{-\theta\widehat{T}_N};\widehat{J}_N=j\mid\widehat{J}_0=k\right]\right) \tag{9.46}$$

then, with $I^\theta=\left(\delta_{jk}\frac{-\nu_{jj}}{-\nu_{jj}+\theta}\right)$,

$$\overline{\gamma}=\zeta\, I^\theta \quad\text{and}\quad \gamma=I^\theta\,\eta \tag{9.47}$$

with, due to (9.47), the results for (ζ,η) being equivalent to those for $(\overline{\gamma},\gamma)$. As a matter of convenience we express some of the remaining results of this section in terms of $(\overline{\gamma},\gamma)$.

Corollary 9.1. *For* $0<z<1$, $\theta>0$, *with* $R=(r_{jk}(\theta))=\left(\delta_{jk}\frac{|y(j)|}{-\nu_{jj}+\theta}\right)$, *we have*

$$\gamma_{00}+(I_{00}-\gamma_{00})\,\overline{\gamma}_{01}\,P_{10}=z\left[I_{00}^\theta\,P_{00}+\gamma_{01}\,I_{11}^\theta\,P_{10}\right] \tag{9.48}$$

$$\gamma_{01}+(I_{00}-\gamma_{00})\,\overline{\gamma}_{01}\,P_{11}=z\left[I_{00}^\theta\,P_{01}+\gamma_{01}\,I_{11}^\theta\,P_{11}\right] \tag{9.49}$$

$$R_{00}\,\overline{\gamma}_{01}\,P_{10}+(I_{00}-\gamma_{00})\,\overline{\gamma}_{01}\,R_{11}\,P_{10}=z\gamma_{01}\,R_{11}\,I_{11}^\theta\,P_{10} \tag{9.50}$$

$$R_{00}\,\overline{\gamma}_{01}\,P_{11}+(I_{00}-\gamma_{00})\,\overline{\gamma}_{01}\,R_{11}\,P_{11}=z\gamma_{01}\,R_{11}\,I_{11}^\theta\,P_{11}. \tag{9.51}$$

Proof. We equate the real parts of the identity (9.43) and put $\omega = 0$. This yields (9.48). We also equate the imaginary parts of (9.43), divide by ω and let $\omega \to 0$. This yields (9.50), in view of (9.48). The proof of (9.49) and (9.51) is similar, starting with the identity (9.44). □

Theorem 9.3 shows that the submatrices $\overline{\chi}_{00}$ and $\overline{\chi}_{01}$ are determined by χ_{00} and χ_{01}. Corollary 9.1 can be used in some important cases to reduce the computation to a single (matrix) equation for $\overline{\gamma}_{01}$, as we will show in the following. Case (a) arises in models with $|E_1| > 1$, while case (b) covers the situation with $|E_1| = 1$. Details of the computations are omitted.

Case (a). If the submatrix P_{11} has an inverse, then

$$\gamma_{00} = zI_{00}^{\theta}\,P_{00} + R_{00}\,\overline{\gamma}_{01}\,R_{11}^{-1}\,P_{10} \tag{9.52}$$

$$\gamma_{01} = zI_{00}^{\theta}\,P_{01} + R_{00}\,\overline{\gamma}_{01}\,R_{11}^{-1}\,P_{11} \tag{9.53}$$

where $\overline{\gamma}_{01}$ satisfies the equation

$$\begin{aligned} &z^2 R_{00}^{-1} I_{00}^{\theta} P_{01} I_{11}^{\theta} + \overline{\gamma}_{01}\left[R_{11}^{-1}P_{10}\right]\overline{\gamma}_{01} \\ &- \left[R_{00}^{-1}\left(I_{00} - zI_{00}^{\theta}P_{00}\right)\overline{\gamma}_{01} + \overline{\gamma}_{01}R_{11}^{-1}\left(I_{11} - zP_{11}I_{11}^{\theta}\right)\right] = 0. \end{aligned}$$

Case (b). If $P_{11} = \mathbf{0}$ and $r_{jj}(\theta) = r_1^{\theta}, j \in E_1$, then

$$\gamma_{00} = zI_{00}^{\theta}\,P_{00} + \frac{1}{r_1^{\theta}} R_{00}\,\overline{\gamma}_{01}\,P_{10} \tag{9.54}$$

$$\gamma_{01} = zI_{00}^{\theta}\,P_{01} \tag{9.55}$$

where $\overline{\gamma}_{01}$ satisfies the equation

$$\begin{aligned} &\left(\overline{\gamma}_{01}\,P_{10}\right)^2 - \left[r_1^{\theta}\,R_{00}^{-1}\left(I_{00} - zI_{00}^{\theta}\,P_{00}\right) + I_{00}\right]\left(\overline{\gamma}_{01}\,P_{10}\right) \\ &\qquad + z^2 r_1^{\theta}\,R_{00}^{-1}\,I_{00}^{\theta}\,P_{01}\,I_{11}^{\theta}\,P_{10} = 0. \end{aligned} \tag{9.56}$$

Example 9.1. Consider the Gaver and Lehoczky [40] model with a single output channel, in which the channel is shared by data and voice calls, with calls having preemptive priority over data. Here $J(t) = 0$ if a call is in progress at time t (i.e. the channel is not available for data transmission), and $J(t) = 1$ otherwise. Thus J has a two state space $\{0, 1\}$ and

$$a(0) = a(1) = c_0, \qquad d(0) = 0,\, d(1) = c_2 \qquad (c_0 < c_2),$$

so that $E_0 = \{0\}$, $E_1 = \{1\}$, $y(0) = c_0$ and $y(1) = c_0 - c_2 = -c_1$. Let the arrival and service rate of calls be denoted by λ and μ respectively, then

$$P = \begin{pmatrix} 0 & 1 \\ 1 & 0 \end{pmatrix} \qquad \text{and} \qquad N = (\nu_{jk}) = \begin{pmatrix} -\mu & \mu \\ \lambda & -\lambda \end{pmatrix}.$$

We may now use (9.47) and (9.54)-(9.56) to conclude that

$$\eta_{01} = z, \qquad \eta_{00} = \frac{\sigma_1}{\sigma_0}\,\zeta_{01}, \qquad \sigma_1\zeta_{01}^2 - \tau(\theta; c_0, c_1)\zeta_{01} + z^2\,\sigma_0 = 0,$$

where $\sigma_0 = \frac{\mu}{c_0}$, $\sigma_1 = \frac{\lambda}{c_1}$, and

$$\tau(\theta; c_0, c_1) = (\sigma_0 + \sigma_1) + \left(\frac{1}{c_0} + \frac{1}{c_1}\right)\theta.$$

This implies that

$$\zeta_{01}(z,\theta) = \frac{\tau(\theta; c_0, c_1) - \sqrt{[\tau(\theta; c_0, c_1)]^2 - 4\,z^2\,\sigma_0\,\sigma_1}}{2\,\sigma_1}. \tag{9.57}$$

For an $M/M/1$ queue with arrival and service rates σ_1 and σ_0 respectively, we denote the busy-period by $\overline{T}^\star$ and the number of customers served during the busy-period by $\overline{N}^\star$. From (9.57) we have the following; see ([85], Section 2.8).

$$\begin{aligned}\zeta_{01}(z,\theta) &= E[z^{\overline{N}} e^{-\theta\overline{T}}; J(\overline{T}) = 1 \mid J(0) = 0] \\ &= E[z^{2\,\overline{N}^\star} e^{-\theta\left(\frac{1}{c_0}+\frac{1}{c_1}\right)\overline{T}^\star}].\end{aligned}$$

Example 9.2. Consider the Virtamo and Norros [116] model, given by (9.6). Denote $c = c_1/c_0$ and $\rho = \lambda/\mu$, in which case $0 < c < 1$ and $\rho > 0$. Equation (9.56) holds for this model and may be proved to be equivalent to:

$$\zeta_{n0}\left[\zeta_{10} - \frac{c\,\mu + \lambda + \mu}{\lambda\,(1-c)}\right] - \frac{z\,c}{\rho\,(1-c)}\,[\zeta_{n-1,0} + \rho\,\zeta_{n+1,0}] = 0, \quad n \geq 1. \tag{9.58}$$

If $J(0) = 1$, it is known from [1] that $\overline{T}$ is equal to the busy-period of an $M/M/1$ queue with arrival rate λ and service rate $c\,\mu$. Thus

$$\zeta_{10}(1,\theta) = \frac{c\,\mu + \lambda + \mu}{2\lambda} - \sqrt{\left(\frac{c\,\mu + \lambda + \mu}{2\lambda}\right)^2 - \frac{c\,\mu}{\lambda}}. \tag{9.59}$$

Using (9.58), the transforms

$$\zeta_{n0}(1,\theta) = (E[e^{-\theta\overline{T}} \mid J(0) = n]), \quad n \geq 1,$$

may be computed recursively, starting with $\zeta_{10}(1,\theta)$ as given in (9.59).

9.4 The Main Results

With the properties for the MRW (T_n, S_n, J_n) established in Sec. 9.3, we are now in a position to derive the main results of the chapter. We first state the following results for the imbedded process (Z_n, I_n, J_n), which follow easily from theorems 7.2 and 7.3.

Lemma 9.4. *If $Z_0 = 0$ a.s. then, for $\theta > 0$ and ω_1, ω_2 real, we have*

$$\left(\sum_{n=0}^{\infty} E\left[e^{-\theta T_n + i\omega_1 Z_n + i\omega_2 I_n}; J_n = k \mid J_0 = j \right] \right)^{-1} = [I - \chi(z, \theta, \omega_1)]\,[I - \overline{\chi}(z, \theta, -\omega_2)]\,. \qquad (9.60)$$

Lemma 9.5. *Suppose that $\mu^\star < 0$. Then $(Z_n, J_n) \overset{\mathcal{L}}{\Longrightarrow} (Z^\star_\infty, J^\star_\infty)$ as $n \to \infty$, for all initial distributions, where $\overset{\mathcal{L}}{\Longrightarrow}$ denotes convergence in distribution, $J^\star_\infty$ is the stationary version of (J_n), and $\left(E\left[e^{i\omega Z^\star_\infty}; J^\star_\infty = k\right]\right)$ is the kth element of the row vector*

$$\pi^\star\ [I - \chi(1, 0, 0)]\ [I - \chi(1, 0, \omega)]$$

where $\pi^\star = (\pi^\star_j,\ j \in E)$.

For finite t, the distribution of $(Z(t), I(t), J(t))$ can be found from Eq. (9.17) using Lemma 9.4. We note that since $\overline{T} = \inf\{t > 0 : Y(t) \leq 0\}$ we have

$$\overline{T} = \inf\{t > 0 : Z(t) = 0 \mid Z(0) = 0\}. \qquad (9.61)$$

Thus $\overline{T}$ is the busy-period of the storage. The transform of $(\overline{T}, J(\overline{T}))$ is $\zeta(1, \theta)$ given by

$$\zeta(1, \theta) = (\zeta_{jk}(1, \theta)) = (E[e^{-\theta \overline{T}}; J(\overline{T}) = k \mid J(0) = j]) \qquad (9.62)$$

In Sec. 9.3 it was shown how this transform can be computed. We recall that $\overline{y}$ is the *net input rate.*

Theorem 9.4. *The busy-period $\overline{T}$ defined by (9.61) has the following properties:*

(a). *Given $J(0) \in E_1$, $\overline{T} = 0$ a.s.*
(b). *If $\overline{y} < 0$ then $\overline{T} < \infty$ a.s.*

Proof. (a). The statement follows immediately from the fact that $y(j) < 0$ for $j \in E_1$.

(b). Since $\overline{y} < 0$ we have $\mu^\star < 0$ (see the proof of Theorem 9.1 (*b*)). This implies that the descending ladder epoch of the MRW (T_n, S_n, J_n) is finite ($\overline{N} < \infty$ a.s.) in virtue of Proposition 7.2. This in turn implies that $T_{\overline{N}} < \infty$ a.s. The statement now follows since $\overline{T} \leq T_{\overline{N}}$. □

The limit behavior of the process $(Z(t), I(t), J(t))$ as $t \to \infty$ can also be obtained from that of the imbedded process (Z_n, I_n, J_n) as $n \to \infty$, by using Lemma 9.5. The following theorems characterize this limit behavior.

Theorem 9.5. *The process $(Z(t), I(t), J(t))$ has the following properties:*
(a). *If $\overline{y} > 0$, then $I(t) \to (Z(0) + m)^- < +\infty$ and $\frac{Z(t)}{t} \to \overline{y}$ a.s.; in particular $Z(t) \to +\infty$ a.s.*
(b). *If $\overline{y} = 0$, then $I(t) \to +\infty$ and $\limsup Z(t) = +\infty$ a.s.*
(c). *If $\overline{y} < 0$, then $\frac{I(t)}{t} \to -\overline{y}$ and $\frac{Z(t)}{t} \to 0$ a.s.; in particular $I(t) \to +\infty$ a.s.*

Proof. (a). We first note that $I(t)$ converges as indicated by Theorem 9.1 (b) and (9.13). The rest of the statement follows directly from Theorem 9.1 (a) and (9.12).

(b). From Theorem 9.1 (c) and (9.13), $I(t) \to (Z(0) + m)^- = +\infty$. Also, from (9.12), since $I(t)$ is nonnegative, $Z(t) \geq Z(0) + Y(t)$. Using Theorem 9.1 (c) we obtain

$$\limsup Z(t) \geq \limsup[Z(0) + Y(t)] = +\infty.$$

(c). Since $Y(t)$ has continuous sample functions and $\frac{Y(t)}{t} \to \overline{y} < 0$, by Theorem 9.1 (a), standard analytical arguments show that

$$\lim \frac{m(t)}{t} = \lim \frac{Y(t)}{t} = \overline{y} < 0.$$

The desired results now follow from (9.13)-(9.15). □

Theorem 9.6. *If $\overline{y} < 0$ then for $z_0, z \geq 0$ and $j, k \in E$, and with $(Z^\star_\infty, J^\star_\infty)$ being the limit distribution of (Z_n, J_n) as given in Lemma 9.5,*

$$\lim_{t\to\infty} P\{Z(t) \leq z; J(t) = k \mid Z(0) = z_0, J(0) = j\} =$$
$$\pi_k \int_{0-}^{\infty} P\{Z^\star_\infty \in dv \mid J^\star_\infty = k\}\, P\{Y_1 \leq z - v \mid J(0) = k\}. \quad (9.63)$$

Proof. Let $N(t) = \sup\{n : T_n \le t\}$. We have

$$\begin{aligned}
&P\{Z(t) \le z; J(t) = k \mid Z(0) = z_0, J(0) = j\} \\
&= \int_{0-}^{\infty} P\left\{Z_{N(t)} \in dv; J_{N(t)} = k \mid Z_0 = z_0, J_0 = j\right\} \\
&\qquad \cdot P\left\{Z(t) \le z \mid Z(T_{N(t)}) = v, J(T_{N(t)}) = k, Z(0) = z_0, J(0) = j\right\} \\
&= P\left\{J_{N(t)} = k \mid J_0 = j\right\} \int_{0-}^{\infty} P\left\{Z_{N(t)} \in dv \mid Z_0 = z_0, J_{N(t)} = k, J_0 = j\right\} \\
&\qquad \cdot P\left\{Z(t) \le z \mid Z(T_{N(t)}) = v, J(T_{N(t)}) = k\right\} \\
&= P\left\{J(t) = k \mid J(0) = j\right\} \int_{0-}^{\infty} P\left\{Z_{N(t)} \in dv \mid Z_0 = z_0, J_{N(t)} = k, J_0 = j\right\} \\
&\qquad \cdot P\left\{[v + y(k)(t - T_{N(t)})]^+ \le z \mid J(T_{N(t)}) = k\right\}
\end{aligned}$$

Since $P\{J(t) = k \mid J(0) = j\} \to \pi_k$ a.s. as $t \to \infty$, the statement follows from the fact that as $t \to \infty$ the following two results hold. Given $J(T_{N(t)}) = k$, $(t - T_{N(t)})$ has by limit distribution T_1, given $J(0) = k$, so that

$$\begin{aligned}
P\left\{[v + y(k)(t - T_{N(t)})]^+ \le z \mid J(T_{N(t)}) = k\right\} \\
\to P\left\{[v + Y_1]^+ \le z \mid J(0) = k\right\} \\
= P\left\{Y_1 \le z - v \mid J(0) = k\right\}.
\end{aligned}$$

Since $\overline{y} < 0$ we have $\mu^\star < 0$, and $N(t) \to \infty$; thus, using Lemma 9.5, we conclude that

$$P\left\{Z_{N(t)} \in dv \mid Z_0 = z_0, J_{N(t)} = k, J_0 = j\right\} \to P\left\{Z^\star_\infty \in dv \mid J^\star_\infty = k\right\}. \square$$

In case $\overline{y} < 0$, we denote by (Z_∞, J_∞) the limit random variable of $(Z, J)(t)$, which, in view of Theorem 9.6, is independent of the initial distribution.

Corollary 9.2. *If $\overline{y} < 0$ we have the following:*

(a). *For $z \ge 0$ we have*

$$P\{Z_\infty \le z \mid J_\infty = k\} = P\{Z^\star_\infty \le z \mid J^\star_\infty = k\} \cdot \left(1 - E\left[e^{-\frac{\nu_{kk}}{y(k)}(Z^\star_\infty - z)} \mid Z^\star_\infty \le z,\ J^\star_\infty = k\right]\right),\ k \in E_0 \tag{9.64}$$

$$P\{Z_\infty > z \mid J_\infty = k\} = P\{Z^\star_\infty > z \mid J^\star_\infty = k\} \cdot \left(1 - E\left[e^{-\frac{\nu_{kk}}{y(k)}(Z^\star_\infty - z)} \mid Z^\star_\infty > z,\ J^\star_\infty = k\right]\right),\ k \in E_1. \tag{9.65}$$

(b). *We have*

$$P\{Z_\infty = 0 \mid J_\infty = k\} = \begin{cases} 0 & k \in E_0 \\ E\left[e^{-\frac{\nu_{kk}}{y(k)} Z^\star_\infty} \mid J^\star_\infty = k\right] & k \in E_1 \end{cases}. \tag{9.66}$$

Proof. (a). Let $z \geq 0$ and $k \in E_0$. From (9.63) we have

$$
\begin{aligned}
&P\{Z_\infty \leq z \mid J_\infty = k\} \\
&\quad = \int_{0-}^{z} P\{Z_\infty^\star \in dv \mid J_\infty^\star = k\}\, P\{Y_1 \leq z - v \mid J(0) = k\} \\
&\quad = \int_{0-}^{z} P\{Z_\infty^\star \in dv \mid J_\infty^\star = k\} \left(1 - e^{\frac{\nu_{kk}}{y(k)}(z-v)}\right) \\
&\quad = P\{Z_\infty^\star \leq z \mid J_\infty^\star = k\} \\
&\qquad \cdot \left(1 - \int_{0-}^{z} P\{Z_\infty^\star \in dv \mid Z_\infty^\star \leq z,\, J_\infty^\star = k\}\, e^{-\frac{\nu_{kk}}{y(k)}(v-z)}\right) \\
&\quad = P\{Z_\infty^\star \leq z \mid J_\infty^\star = k\} \left(1 - E\left[e^{-\frac{\nu_{kk}}{y(k)}(Z_\infty^\star - z)} \mid Z_\infty^\star \leq z,\, J_\infty^\star = k\right]\right).
\end{aligned}
$$

This gives (9.64). The proof of (9.65) is similar.
(b). The statement follows from the fact that using (9.63) we have

$$
\begin{aligned}
&P\{Z_\infty = 0 \mid J_\infty = k\} \\
&\quad = \int_{0-}^{\infty} P\{Z_\infty^\star \in dv \mid J_\infty^\star = k\}\, P\{Y_1 \leq -v \mid J(0) = k\}. \qquad \square
\end{aligned}
$$

We denote by $I_k(t)$ the unsatisfied demand in state k during $(0, t]$, so that

$$
\begin{aligned}
I_k(t) &= \int_0^t y(J(s))^- \mathbf{1}_{\{Z(s)=0,\, J(s)=k\}} ds \\
&= y(k)^- \int_0^t \mathbf{1}_{\{Z(s)=0,\, J(s)=k\}} ds. \qquad (9.67)
\end{aligned}
$$

If $k \in E_0$, then $y(k)^- = 0$ and $I_k(t) = 0$. If $k \in E_1$, we have the following important result for the performance analysis of the system.

Corollary 9.3. *If $\overline{y} < 0$ and $k \in E_1$, then*

$$
\lim_{t\to\infty} \frac{I_k(t)}{t} = -y(k)\, \pi_k\, E\left[e^{-\frac{\nu_{kk}}{y(k)} Z_\infty^\star} \mid J_\infty^\star = k\right] \qquad (9.68)
$$

and

$$
\lim_{t\to\infty} \frac{I_k(t)}{I(t)} = \frac{y(k)\, \pi_k\, E\left[e^{-\frac{\nu_{kk}}{y(k)} Z_\infty^\star} \mid J_\infty^\star = k\right]}{\sum_{j \in E_1} y(j)\, \pi_j\, E\left[e^{-\frac{\nu_{jj}}{y(j)} Z_\infty^\star} \mid J_\infty^\star = j\right]}. \qquad (9.69)
$$

Proof. Using Theorem 9.6 we conclude that

$$
\frac{1}{t} \int_0^t \mathbf{1}_{\{Z(s)=0, J(s)=k\}} ds \longrightarrow P\{Z_\infty = 0,\, J_\infty = k\}.
$$

This implies (9.68) in view of (9.66)-(9.67). Also, since $I(t) = \sum_{j \in E_1} I_j(t)$, (9.69) follows by using (9.68). $\square$

Example 9.1 (Continuation) We note that $\pi_0^\star = \pi_1^\star = 1/2, \zeta_{01}(1,0) = 1, \eta_{01}(1,0) = 1$ and $\eta_{00}(1,0) = \rho$, with $\rho = \sigma_1/\sigma_0$. We now assume that $\rho < 1$. Since

$$\pi^\star \, [I - \chi(1,0,0)] \, [I - \chi(1,0,\omega)]^{-1} = \frac{1}{2}\left[(1-\rho) + \rho\frac{\sigma_0 - \sigma_1}{(\sigma_0 - \sigma_1) - i\omega} \qquad \frac{\sigma_0 - \sigma_1}{(\sigma_0 - \sigma_1) - i\omega}\right],$$

we conclude from Lemma 9.5 that $P\{Z_\infty^\star > z \mid J_\infty^\star = 0\} = \rho\, e^{-(\sigma_0 - \sigma_1)z}$ for $z \geq 0$, and similarly $P\{Z_\infty^\star > z \mid J_\infty^\star = 1\} = e^{-(\sigma_0 - \sigma_1)z}$. With $\pi_0 = \lambda/(\lambda + \mu)$ and $\pi_1 = \mu/(\lambda + \mu)$ we conclude, using Theorem 9.6, that for $z \geq 0$,

$$\begin{cases} P\{Z_\infty > z; J_\infty = 0\} = \pi_0 \, e^{-(\sigma_0 - \sigma_1)z} \\ P\{Z_\infty > z; J_\infty = 1\} = \pi_1 \, \rho \, e^{-(\sigma_0 - \sigma_1)z} \end{cases}. \tag{9.70}$$

Finally, using Corollary 9.3, we conclude that a.s.

$$I_0(t) = 0, \quad \forall t \qquad \text{and} \qquad \lim_{t \to \infty} \frac{I_1(t)}{t} = c_1 \, \pi_1 \, (1 - \rho).$$

We note that this example has been considered also by Chen and Yao [25], Gaver and Miller [41], and Kella and Whitt [53], in the context of storage models for which the net input is alternatingly nonincreasing and nondecreasing, and by Karandikar and Kulkarni ([49], Case 1 of Example 1 – Section 6) with the storage level being a particular case of a Markov-modulated reflected Brownian motion.

Bibliography

[1] Aalto, S. (1998). Characterization of the output rate process for a Markovian storage model, *J. Appl. Probab.* **35**, 1, pp. 184–199.

[2] Anick, D., Mitra, D. and Sondhi, M. M. (1982). Stochastic theory of a data-handling system with multiple sources, *Bell Syst. Tech. J.* **61**, 8, pp. 1871–1894.

[3] Anisimov, V. V. (1970). Certain theorems on limit distributions of sums of random variables that are connected in a homogeneous Markov chain, *Dopovīdī Akad. Nauk Ukraïn. RSR Ser. A* **1970**, pp. 99–103, 187.

[4] Antunes, N., Nunes, C. and Pacheco, A. (2005). Functionals of Markovian branching D-BMAPs, *Stoch. Models* **21**, 2-3, pp. 261–278.

[5] Arjas, E. (1972). On a fundamental identity in the theory of semi-Markov processes, *Advances in Appl. Probability* **4**, pp. 258–270.

[6] Arjas, E. and Speed, T. P. (1973a). Symmetric Wiener-Hopf factorisations in Markov additive processes, *Z. Wahrscheinlichkeitstheorie und Verw. Gebiete* **26**, pp. 105–118.

[7] Arjas, E. and Speed, T. P. (1973b). Topics in Markov additive processes, *Math. Scand.* **33**, pp. 171–192.

[8] Arndt, K. (1985). On the distribution of the supremum of a random walk on a Markov chain, in A. A. Borovkov and A. Balakrishnan (eds.), *Advances in Probability Theory: Limit Theorems for Sums of Random Variables*, Translations Series in Mathematics and Engineering (Optimization Software, New York), pp. 253–267.

[9] Asmussen, S. (1987). *Applied Probability and Queues* (Wiley, New York).

[10] Asmussen, S. (1991). Ladder heights and the Markov-modulated $M/G/1$ queue, *Stochastic Process. Appl.* **37**, 2, pp. 313–326.

[11] Asmussen, S. (1995). Stationary distributions for fluid flow models with or without Brownian noise, *Comm. Statist. Stochastic Models* **11**, 1, pp. 21–49.

[12] Asmussen, S. (2003). *Applied Probability and Queues*, 2nd edn., Applications of Mathematics, Stochastic Modelling and Applied Probability, Vol. 51 (Springer-Verlag, New York).

[13] Asmussen, S. and Koole, G. (1993). Marked point processes as limits of

Markovian arrival streams, *J. Appl. Probab.* **30**, 2, pp. 365–372.
[14] Billingsley, P. (1995). *Probability and Measure*, 3rd edn., Wiley Series in Probability and Mathematical Statistics (John Wiley & Sons Inc., New York).
[15] Blondia, C. (1993). A discrete-time batch Markovian arrival process as B-ISDN traffic model, *Belg. J. Oper. Res. Statist. Comput. Sci.* **32**, pp. 3–23.
[16] Blondia, C. and Casals, O. (1992). Statistical multiplexing of VBR sources: A matrix-analytic approach, *Perform. Evaluation* **16**, 1-3, pp. 5–20.
[17] Blumenthal, R. M. (1992). *Excursions of Markov Processes*, Probability and its Applications (Birkhäuser Boston Inc., Boston, MA).
[18] Blumenthal, R. M. and Getoor, R. K. (1968). *Markov Processes and Potential Theory*, Pure and Applied Mathematics, Vol. 29 (Academic Press, New York).
[19] Bondesson, L. (1981). On occupation times for quasi-Markov processes, *J. Appl. Probab.* **18**, 1, pp. 297–301.
[20] Borisov, A. V., Korolev, V. Y. and Stefanovich, A. I. (2006). Hidden Markov models of plasma turbulence, in *Stochastic Models of Structural Plasma Turbulence*, Mod. Probab. Stat. (VSP, Leiden), pp. 345–400.
[21] Breuer, L. (2002). On Markov-additive jump processes, *Queueing Syst.* **40**, 1, pp. 75–91.
[22] Breuer, L. (2003). *From Markov Jump Processes to Spatial Queues* (Kluwer Academic Publishers, Dordrecht).
[23] Browne, S. and Zipkin, P. (1991). Inventory models with continuous, stochastic demands, *Ann. Appl. Probab.* **1**, 3, pp. 419–435.
[24] Chandrasekhar, S. (1943). Stochastic problems in physics and astronomy, *Rev. Modern Phys.* **15**, pp. 1–89.
[25] Chen, H. and Yao, D. D. (1992). A fluid model for systems with random disruptions, *Oper. Res.* **40**, suppl. 2, pp. S239–S247.
[26] Chung, K. L. (1967). *Markov Chains with Stationary Transition Probabilities*, Second edition. Die Grundlehren der mathematischen Wissenschaften, Band 104 (Springer-Verlag New York, Inc., New York).
[27] Çinlar, E. (1969). Markov renewal theory, *Advances in Appl. Probability* **1**, pp. 123–187.
[28] Çinlar, E. (1972a). Markov additive processes. I, *Z. Wahrscheinlichkeitstheorie und Verw. Gebiete* **24**, pp. 85–93.
[29] Çinlar, E. (1972b). Markov additive processes. II, *Z. Wahrscheinlichkeitstheorie und Verw. Gebiete* **24**, pp. 95–121.
[30] Çinlar, E. (1975). *Introduction to Stochastic Processes* (Prentice Hall, Englewood Cliffs, NJ).
[31] Disney, R. L. and Kiessler, P. C. (1987). *Traffic Processes in Queueing Networks* (The Johns Hopkins University Press, Baltimore, MD).
[32] Elliott, R. J. and Swishchuk, A. V. (2007). Pricing options and variance swaps in Markov-modulated Brownian Markets, in R. S. Mamon and R. J. Elliott (eds.), *Hidden Markov Models in Finance*, International Series in Operations Research & Management Science, Vol. 104 (Springer, New York), pp. 45–68.

[33] Elwalid, A. and Mitra, D. (1991). Analysis and design of rate-based congestion control of high speed networks, I: Stochastic fluid models, access regulation, *Queueing Systems Theory Appl.* **9**, 1-2, pp. 29–63.

[34] Feller, W. (1949). Fluctuation theory of recurrent events, *Trans. Amer. Math. Soc.* **67**, pp. 98–119.

[35] Feller, W. (1968). *An Introduction to Probability Theory and its Applications*, Vol. 1, 3rd edn. (John Wiley, New York).

[36] Feller, W. (1971). *An Introduction to Probability Theory and its Applications. Vol. II*, 2nd edn. (John Wiley & Sons Inc., New York).

[37] Fischer, W. and Meier-Hellstern, K. S. (1993). The Markov-modulated Poisson process (MMPP) cookbook, *Perform. Eval.* **18**, 2, pp. 149–171.

[38] Friedman, A. (1982). *Foundations of Modern Analysis* (Dover Publications Inc., New York).

[39] Fristedt, B. (1974). Sample functions of stochastic processes with stationary, independent increments, in P. Ney (ed.), *Advances in Probability and Related Topics*, Vol. 3 (Marcel Dekker, New York), pp. 241–396.

[40] Gaver, D. and Lehoczky, J. (1982). Channels that cooperatively service a data stream and voice messages, *IEEE Trans. Commun.* **30**, 5, pp. 1153–1162.

[41] Gaver, D. P., Jr. and Miller, R. G., Jr. (1962). Limiting distributions for some storage problems, in *Studies in Applied Probability and Management Science* (Stanford Univ. Press, Stanford, Calif.), pp. 110–126.

[42] Goldstein, J. A. (1985). *Semigroups of Linear Operators and Applications*, Oxford Mathematical Monographs (The Clarendon Press Oxford University Press, New York).

[43] He, Q.-M. and Neuts, M. F. (1998). Markov chains with marked transitions, *Stochastic Process. Appl.* **74**, 1, pp. 37–52.

[44] Heyde, C. C. (1967). A limit theorem for random walks with drift, *J. Appl. Probability* **4**, pp. 144–150.

[45] Janssen, J. and Manca, R. (2006). *Applied Semi-Markov Processes* (Springer, New York).

[46] Janssen, J. and Manca, R. (2007). *Semi-Markov Risk Models for Finance, Insurance and Reliability* (Springer, New York).

[47] Kac, M. (1947). Random walk and the theory of Brownian motion, *Amer. Math. Monthly* **54**, pp. 369–391.

[48] Kac, M. (1964). Probability, *Sci. Amer.* **211**, pp. 92–108.

[49] Karandikar, R. L. and Kulkarni, V. G. (1995). Second-order fluid-flow models - reflected Brownian motion in a random environment, *Oper. Res.* **43**, 1, pp. 77–88.

[50] Karatzas, I. and Shreve, S. E. (1988). *Brownian Motion and Stochastic Calculus*, Graduate Texts in Mathematics, Vol. 113 (Springer-Verlag, New York).

[51] Karr, A. F. (1991). *Point Processes and Their Statistical Inference*, 2nd edn., Probability: Pure and Applied, Vol. 7 (Marcel Dekker Inc., New York).

[52] Keilson, J. and Wishart, D. M. G. (1964). A central limit theorem for

processes defined on a finite Markov chain, *Proc. Cambridge Philos. Soc.* **60**, pp. 547–567.

[53] Kella, O. and Whitt, W. (1992). A storage model with a two-state random environment, *Oper. Res.* **40**, suppl. 2, pp. S257–S262.

[54] Kingman, J. F. C. (1965). Linked systems of regenerative events, *Proc. London Math. Soc. (3)* **15**, pp. 125–150.

[55] Kingman, J. F. C. (1972). *Regenerative Phenomena*, Wiley Series in Probability and Mathematical Statistics (John Wiley, London).

[56] Kingman, J. F. C. (1993). *Poisson Processes*, Oxford Studies in Probability, Vol. 3 (The Clarendon Press Oxford University Press, New York).

[57] Kulkarni, V. G. (1995). *Modeling and Analysis of Stochastic Systems*, Texts in Statistical Science Series (Chapman and Hall Ltd., London).

[58] Kulkarni, V. G. and Prabhu, N. U. (1984). A fluctuation theory for Markov chains, *Stochastic Process. Appl.* **16**, 1, pp. 39–54.

[59] Lalam, N. (2007). Statistical inference for quantitative polymerase chain reaction using a hidden Markov model: a Bayesian approach, *Stat. Appl. Genet. Mol. Biol.* **6**, pp. Art. 10, 35 pp. (electronic).

[60] Limnios, N. and Oprişan, G. (2001). *Semi-Markov Processes and Reliability*, Statistics for Industry and Technology (Birkhäuser Boston Inc., Boston, MA).

[61] Lotov, V. I. and Orlova, N. G. (2006). Asymptotic expansions for the distribution of the number of crossings of a strip by a random walk defined on a Markov chain, *Sibirsk. Mat. Zh.* **47**, 6, pp. 1303–1322.

[62] Lucantoni, D. M. (1991). New results on the single server queue with a batch Markovian arrival process, *Comm. Statist. Stochastic Models* **7**, 1, pp. 1–46.

[63] Lucantoni, D. M. (1993). The BMAP/G/1 queue: A tutorial, in L. Donatiello and R. D. Nelson (eds.), *Performance/SIGMETRICS Tutorials*, Lecture Notes in Computer Science, Vol. 729 (Springer), pp. 330–358.

[64] Lucantoni, D. M., Meier-Hellstern, K. S. and Neuts, M. F. (1990). A single-server queue with server vacations and a class of nonrenewal arrival processes, *Adv. in Appl. Probab.* **22**, 3, pp. 676–705.

[65] Macci, C. (2005). Random sampling for continuous time Markov additive processes, *Publ. Mat. Urug.* **10**, pp. 77–89.

[66] Machihara, F. (1988). On the overflow processes from the $\mathrm{PH}_1 + PH_2/M/S/K$ queue with two independent PH-renewal inputs, *Perform. Eval.* **8**, 4, pp. 243–253.

[67] Maisonneuve, B. (1971). Ensembles régénératifs, temps locaux et subordinateurs, in *Séminaire de Probabilités, V*, Lecture Notes in Mathematics, Vol. 191 (Springer, Berlin), pp. 147–169.

[68] Miller, H. D. (1962a). Absorption probabilities for sums of random variables defined on a finite Markov chain, *Proc. Cambridge Philos. Soc.* **58**, pp. 286–298.

[69] Miller, H. D. (1962b). A matrix factorization problem in the theory of random variables defined on a finite Markov chain, *Proc. Cambridge Philos. Soc.* **58**, pp. 268–285.

[70] Mitra, D. (1988). Stochastic theory of a fluid model of producers and consumers coupled by a buffer, *Adv. in Appl. Probab.* **20**, 3, pp. 646–676.

[71] Narayana, S. and Neuts, M. F. (1992). The first two moment matrices of the counts for the Markovian arrival process, *Comm. Statist. Stochastic Models* **8**, 3, pp. 459–477.

[72] Neuts, M. F. (1989). *Structured Stochastic Matrices of $M/G/1$ Type and Their Applications*, Probability: Pure and Applied, Vol. 5 (Marcel Dekker Inc., New York).

[73] Neuts, M. F. (1992). Models based on the Markovian arrival process, *IEICE Trans. Commun.* **E75-B**, pp. 1255–65.

[74] Neveu, J. (1961). Une generalisation des processus à accroissements positifs independants, *Abh. Math. Sem. Univ. Hamburg* **25**, pp. 36–61.

[75] Newbould, M. (1973). A classification of a random walk defined on a finite Markov chain, *Z. Wahrscheinlichkeitstheorie und Verw. Gebiete* **26**, pp. 95–104.

[76] Nielsen, B. F., Nilsson, L. A. F., Thygesen, U. H. and Beyer, J. E. (2007). Higher order moments and conditional asymptotics of the batch Markovian arrival process, *Stoch. Models* **23**, 1, pp. 1–26.

[77] Nogueira, A., Salvador, P., Valadas, R. and Pacheco, A. (2003). Modeling network traffic with multifractal behavior, *Telecommunication Systems* **24**, 2-4, pp. 339–362.

[78] Nogueira, A., Salvador, P., Valadas, R. and Pacheco, A. (2004). Fitting self-similar traffic by a superposition of MMPPs modeling the distribution at multiple time scales, *IEICE Trans. Commun.* **E87B**, 3, pp. 678–688.

[79] Oliveira, D. C., da Silva, C. Q. and Chaves, L. M. (2006). Hidden Markov models applied to DNA sequence analysis, *Rev. Mat. Estatíst.* **24**, 2, pp. 51–66.

[80] Olivier, C. and Walrand, J. (1994). On the existence of finite-dimensional filters for Markov-modulated traffic, *J. Appl. Probab.* **31**, 2, pp. 515–525.

[81] Pacheco, A. and Prabhu, N. U. (1995). Markov-additive processes of arrivals, in J. H. Dshalalow (ed.), *Advances in Queueing: Theory, Methods, and Open Problems*, Probability and Stochastics Series (CRC, Boca Raton, FL), pp. 167–194.

[82] Pacheco, A. and Prabhu, N. U. (1996). A Markovian storage model, *Ann. Appl. Probab.* **6**, 1, pp. 76–91.

[83] Palm, C. (1943). Intensitätsschwankungen im Fernsprechverkehr, *Ericsson Technics* **44**, pp. 1–189.

[84] Percus, J. K. and Percus, O. E. (2006). The statistics of words on rings, *Comm. Pure Appl. Math.* **59**, 1, pp. 145–160.

[85] Prabhu, N. U. (1965). *Queues and Inventories. A Study of Their Basic Stochastic Processes* (John Wiley & Sons Inc., New York).

[86] Prabhu, N. U. (1980). *Stochastic Storage Processes: Queues, Insurance Risk, and Dams*, Applications of Mathematics, Vol. 15 (Springer-Verlag, New York).

[87] Prabhu, N. U. (1985). Wiener-Hopf factorization of Markov semigroups. I. The countable state space case, in M. Iosifescu (ed.), *Proceedings of the*

Seventh Conference on Probability Theory (Braşov, 1982) (VNU Sci. Press, Utrecht), pp. 315–324.

[88] Prabhu, N. U. (1988). Theory of semiregenerative phenomena, *J. Appl. Probab.* **25A**, pp. 257–274.

[89] Prabhu, N. U. (1991). Markov-renewal and Markov-additive processes – a review and some new results, in *Analysis and Geometry 1991* (Korea Adv. Inst. Sci. Tech., Taejŏn), pp. 57–94.

[90] Prabhu, N. U. (1994). Further results for semiregenerative phenomena, *Acta Appl. Math.* **34**, 1-2, pp. 213–223.

[91] Prabhu, N. U. (1998). *Stochastic Storage Processes: Queues, Insurance Risk, Dams, and Data Communication*, 2nd edn., Applications of Mathematics, Vol. 15 (Springer-Verlag, New York).

[92] Prabhu, N. U. (2007). *Stochastic Processes: Basic Theory and Its Applications* (World Scientific, Hackensack, NJ).

[93] Prabhu, N. U. and Pacheco, A. (1995). A storage model for data communication systems, *Queueing Systems Theory Appl.* **19**, 1-2, pp. 1–40.

[94] Prabhu, N. U. and Tang, L. C. (1994). Markov-modulated single-server queueing systems, *J. Appl. Probab.* **31A**, pp. 169–184.

[95] Prabhu, N. U., Tang, L. C. and Zhu, Y. (1991). Some new results for the Markov random walk, *J. Math. Phys. Sci.* **25**, 5-6, pp. 635–663.

[96] Prabhu, N. U. and Zhu, Y. (1989). Markov-modulated queueing systems, *Queueing Systems Theory Appl.* **5**, 1-3, pp. 215–245.

[97] Presman, È. L. (1969). Factorization methods, and a boundary value problem for sums of random variables given on a Markov chain, *Izv. Akad. Nauk SSSR Ser. Mat.* **33**, pp. 861–900.

[98] Resnick, S. I. (1999). *A Probability Path* (Birkhäuser Boston Inc., Boston, MA).

[99] Rogers, L. C. G. (1994). Fluid models in queueing theory and Wiener-Hopf factorization of Markov chains, *Ann. Appl. Probab.* **4**, 2, pp. 390–413.

[100] Rogers, L. C. G. and Shi, Z. (1994). Computing the invariant law of a fluid model, *J. Appl. Probab.* **31**, 4, pp. 885–896.

[101] Rudemo, M. (1973). Point processes generated by transitions of Markov chains, *Advances in Appl. Probability* **5**, pp. 262–286.

[102] Salvador, P., Pacheco, A. and Valadas, R. (2004). Modeling ip traffic: joint characterization of packet arrivals and packet sizes using BMAPs, *Computer Networks* **44**, 3, pp. 335–352.

[103] Salvador, P., Valadas, R. T. and Pacheco, A. (2003). Multiscale fitting procedure using markov modulated Poisson processes, *Telecommunication Systems* **23**, 1-2, pp. 123–148.

[104] Serfozo, R. F. (1993). Queueing networks with dependent nodes and concurrent movements, *Queueing Systems Theory Appl.* **13**, 1-3, pp. 143–182.

[105] Sevast′yanov, B. A. (1962). Influence of a storage bin capacity on the average standstill time of a production line, *Theor. Probability Appl.* **7**, 4, pp. 429–438.

[106] Siu, T. K. (2007). On fair valuation of participating life insurance policies with regime switching, in R. S. Mamon and R. J. Elliott (eds.), *Hidden*

Markov Models in Finance, International Series in Operations Research & Management Science, Vol. 104 (Springer, New York), pp. 31–43.

[107] Smith, W. L. (1955). Regenerative stochastic processes, *Proc. Roy. Soc. London. Ser. A.* **232**, pp. 6–31.

[108] Smith, W. L. (1960). Infinitesimal renewal processes, in I. Olkin, S. G. Ghurye, W. Hoeffding, W. G. Madow and H. B. Mann (eds.), *Contributions to Probability and Statistics: Essays in Honour of Harold Hoteling*, Stanford Studies in Mathematics and Statistics, Vol. 2 (Stanford University Press, Stanford, California), pp. 396–413.

[109] Takács, L. (1976). On fluctuation problems in the theory of queues, *Advances in Appl. Probability* **8**, 3, pp. 548–583.

[110] Takács, L. (1978). On fluctuations of sums of random variables, in *Studies in Probability and Ergodic Theory*, Adv. in Math. Suppl. Stud., Vol. 2 (Academic Press, New York), pp. 45–93.

[111] Takine, T., Sengupta, B. and Hasegawa, T. (1994). A conformance measure for traffic shaping in high-speed networks with an application to the leaky bucket, in *INFOCOM*, pp. 474–481.

[112] Tang, L. C. (1993). Limit theorems for Markov random walks, *Statist. Probab. Lett.* **18**, 4, pp. 265–270.

[113] Temperley, D. (2007). *Music and Probability* (MIT Press, Cambridge, MA).

[114] van Doorn, E. A., Jagers, A. A. and de Wit, J. S. J. (1988). A fluid reservoir regulated by a birth-death process, *Comm. Statist. Stochastic Models* **4**, 3, pp. 457–472.

[115] van Doorn, E. A. and Regterschot, G. J. K. (1988). Conditional PASTA, *Oper. Res. Lett.* **7**, 5, pp. 229–232.

[116] Virtamo, J. and Norros, I. (1994). Fluid queue driven by an $M/M/1$ queue, *Queueing Systems Theory Appl.* **16**, 3-4, pp. 373–386.

[117] Xing, X.-y. and Wang, X.-q. (2008). On the asymptotic behavior of the storage process fed by a Markov modulated Brownian motion, *Acta Math. Appl. Sin. Engl. Ser.* **24**, 1, pp. 75–84.

[118] Yamada, H. and Machihara, F. (1992). Performance analysis of a statistical multiplexer with control on input and/or service processes, *Perform. Eval.* **14**, 1, pp. 21–41.

Index

additive
  additive component, 8, 23, 24, 34, 52, 78, 81, 83, 91, 93, 94, 97, 103, 104, 105, 109, 124, 158, 174, 177
  additive functional, 94, 133, 163, 164, 165, 166, 167, 189, 193
  additive process, 5, 7, 8, 9, 11, 13, 15, 17, 19, 21, 23, 24, 28, 29, 31, 32, 36, 40, 42, 43, 49, 56, 57, 61, 103, 108, 129, 131, 135, 139, 155, 190
aperiodic/aperiodicity, 75, 184, 185, 186
arrival
  arrival component, 105, 112, 113, 114, 115
  arrival epoch, 105, 108, 112, 116, 117, 134, 137, 139
  arrival rate, 123, 137, 151, 151, 177, 180, 191, 200
  batch arrival, 108, 137
  classes of arrivals, 8, 103, 104, 114
  customer arrival, 134, 137, 141
  non-recorded arrival, 140
  phase-type arrival, 107, 108, 114
  recorded arrival, 140
arrival process
  batch Markovian arrival process (BMAP), 108
  counting (arrival) process, 9, 86, 137, 138, 139
  hyperexponential arrival process, 107
  Markovian arrival process, 107, 108
  phase-type arrival process (CPH), 112
  Yamada and Machihara's arrival process, 108, 116, 117, 118
assumptions, 4, 25, 26, 28, 34, 51, 155, 156, 157, 158

Banach space, 114, 175, 177
BMAP (see arrival process)
birth
  birth process, 23
  birth rate, 157
  birth-death process, 213
Brownian
  Brownian motion, 4, 8, 192, 205
  Markov-modulated Brownian motion, 8
  reflected Brownian motion, 192, 205

compactification, 22, 51
conditional
  conditional distribution, 23, 24, 36, 104, 195
  conditional independence, 11
  conditional probability, 10, 44, 51
convergence
  convergence in distribution, 124, 150

convergence theorem, 15, 72, 84, 85, 86
covariance/covariance matrix, 122, 124
CPH (see arrival process)
CPH-MRP (see arrival process)

data
data communication, 108, 155, 157, 190, 191
data-handling, 155, 191
degeneracy, 62, 67, 68, 69, 71, 81, 142
demand
demand process, 164, 165, 167, 172, 182
demand rate, 157, 159, 160, 174, 184
demand rejection rate, 180, 188
unsatisfied demand, 159, 160, 163, 166, 180, 182, 183, 189, 192, 194, 204
unsatisfied demand process, 182
distribution
conditional distribution, 23, 24, 36, 104, 195
convergence in distribution, 124, 150
equality in distribution, 69, 95
exponential distribution, 23, 151, 153, 155, 156, 191
finite-dimensional distribution, 43
limit distribution, 30, 100, 146, 147, 148, 150, 185, 186, 202, 203
normal distribution, 98, 124, 150
probability distribution, 73, 73, 83, 83

ergodic/ergodicity, 62, 63, 68, 75, 94, 123, 160, 164, 169, 170, 184, 185, 186, 195
environment, 7, 24, 101, 191
equation
Chapman-Kolmogorov equation, 10, 24, 25, 28, 34, 64, 85, 111, 115
differential equation, 28, 29, 32, 34, 115, 118, 120, 121
functional equation, 178, 179, 180
integral equation, 12, 13, 14, 15, 19, 21, 53, 85, 158, 159, 161, 189, 193

factorization
basic factorization identity, 89
Wiener-Hopf factorization, 61, 62, 63, 72, 87, 89, 142, 189, 192, 196
filtration, 161, 163, 175
finance, 8, 62, 93, 94, 100
fluctuation
fluctuation behavior, 49, 61, 63, 91, 194, 195
fluctuation theory, 49, 61, 62, 63, 78, 87
fluid
fluid input, 156
fluid model, 156
fluid source, 157
function
bounded (real) function, 36, 51, 85, 159
centering function, 37
continuous function, 26, 58
characteristic function, 153
density function, 153
distribution function, 19, 34
drift function, 166, 174
indicator function, 3, 4
generating function, 28, 29, 32, 91, 106, 107, 118, 119
measurable function, 160, 164
maximum function, 48, 49, 72, 144
nonnegative function, 19, 132, 133, 189
p-function, 4, 52
probability function, 4, 114, 135
renewal function, 19, 21
sample function, 5, 17, 19, 112, 115, 163, 191, 202

functional
additive functional, 94, 133, 163, 164, 165, 166, 167, 189, 193
functional equation, 178, 179, 180
maximum functional, 48, 72, 144
minimum functional, 49, 78, 143, 144
extremum functional, 75, 77, 145, 146

generator
generator matrix, 25, 29, 35, 59, 107, 110, 111, 112, 113, 115, 116, 129, 133, 140, 192
generator of a contraction semigroup, 178
generator of a semigroup, 178
infinitesimal generator, 8, 26, 36, 51, 55, 59, 110, 112, 113, 115, 159, 160, 165, 166, 169, 170, 171, 172, 173, 176
geometric increase, 130

heavy traffic, 100
hidden
hidden Markov chain, 65
hidden Markov model (HMM), 62, 65
HMM (see hidden Markov model)
homogeneous
homogeneous case, 24, 64, 124
homogeneous MAP, 104, 124, 132
homogeneous Markov chain, 1, 3, 42, 48
homogeneous Markov process, 4, 50, 93, 126
homogeneous Markov subordinator, 132
homogeneous MRW, 127, 143
homogeneous process, 10, 132

identity
basic factorization identity, 89
Miller-Kemperman identity, 61, 63
imbedded/imbedding, 2, 12, 21, 39, 61, 62, 66, 68, 69, 72, 81, 84, 87, 127, 128, 186, 189, 192, 193, 195, 201, 202
increments
independent increments, 8, 23, 24, 106
normalized increments, 94
stationary and independent increments, 106
input
batch input, 112
cumulative input, 155
fluid input, 156, 156
input component, 108, 112, 131, 158, 174, 192
input process, 108, 158, 160, 166, 189, 190, 191, 194, 197
input rate, 155, 156, 157, 159, 191, 194, 201
Markov-additive input, 108
total input, 135, 158
intensity
intensity matrix, 160, 165, 170, 173, 176, 185
intensity rate, 106
irreducible/irreducibility, 63, 68, 69, 72, 81, 113, 121, 123, 124, 143, 184, 185, 186

ladder
ladder epoch, 49, 196, 197, 202
ladder height, 63, 145, 196
ladder measure, 143
ladder sets, 79, 141, 142
ladder variable, 196
law
invariant law, 192
law of iterated logarithm, 93, 94, 97, 99
law of large numbers, 82, 83, 94, 114, 123, 124
stable law, 150
limit
central limit theorem, 93, 94, 97, 98, 100, 124, 150
Césaro limit, 81
limit behavior, 21, 66, 79, 202

limit demand, 188
limit distribution, 30, 100, 146, 147, 148, 150, 185, 186, 202, 203
limit probability, 188
limit random variable, 203
limit rate, 188
limit result, 63, 150
limit theorems, 13, 21, 83, 93, 94, 95, 97, 99, 101, 103

linear
linear combination, 103, 119, 126, 127, 129, 130
linear function, 160, 177
linear operator, 114, 177
linear transformation, 103, 125, 126, 127, 128

MAP (see Markov-additive process)

Markov
Markov chain, 1, 3, 8, 9, 11, 24, 39, 40, 42, 48, 49, 54, 55, 57, 59, 61, 62, 65, 66, 68, 72, 75, 82, 94, 99, 100, 105, 106, 107, 110, 112, 123, 133, 140, 141, 142, 143, 155, 157, 158, 160, 164, 170, 173, 174, 176, 177, 184, 185, 186, 189, 191, 192
Markov component, 5, 7, 23, 24, 105, 106, 107, 108, 109, 110, 111, 113, 114, 117, 118, 124, 130, 131, 133
Markov process, 1, 4, 7, 8, 9, 10, 11, 22, 23, 24, 29, 30, 32, 34, 40, 50, 62, 64, 93, 103, 104, 106, 116, 117, 125, 126, 131, 156, 163, 166, 167
Markov property, 12, 42, 48, 66, 68, 75, 79, 130
Markov random walk (MRW), 93, 141, 192
Markov renewal customers, 112
Markov renewal equation, 9, 19, 20, 21, 22
Markov renewal measure, 51, 52, 76, 145
Markov renewal process, 7, 8, 9, 10, 16, 39, 42, 43, 62, 63, 108, 117, 142
Markov renewal property, 31, 31, 33, 33, 44, 44, 47, 47
Markov renewal theory, 9, 39, 43, 57
Markov source, 62, 65
Markov subordinator, 8, 39, 40, 42, 50, 54, 55, 105, 106, 115, 126, 132
Markov transition, 4

Markovian
batch Markovian arrival process (BMAP), 108
Markovian arrival process, 107
Markovian nature, 104
Markovian network, 111
Markovian storage model, 189, 191, 193, 195, 197, 199, 201, 203, 205

Markov-additive
MAP of arrivals, 103, 104, 105, 106, 107, 108, 109, 110, 111, 112, 113, 114, 115, 116, 117, 118, 119, 121, 124, 127, 128, 129, 130, 131, 132, 133, 134, 135, 136, 137, 138, 139, 140
Markov-additive input, 108
Markov-additive process (MAP), 40, 155
Markov-additive property, 53, 103, 124, 133
Markov-additive structure, 109

Markov-compound Poisson process (MCPP), 159

Markov-modulated
Markov-modulated Brownian motion, 8
Markov-modulated case, 142
Markov-modulated experiment, 65
Markov-modulated fluid, 157
Markov-modulated jump, 34, 36, 50, 52

- Markov-modulated measure, 72, 76, 93, 99, 111, 125
- Markov-modulated M/G/1, 141, 180, 187
- Markov-modulated M/M/1, 141, 143, 151, 152
- Markov-modulated M/M/1/K, 136, 137
- Markov-modulated process, 7, 8
- Markov-modulated Poisson process (MMPP), 209
- Markov-modulated queue, 24, 62, 141, 144
- Markov-modulated single server queue, 141, 143, 145, 147, 149, 151, 153
- Markov-modulated transition, 107

Markov-Bernoulli
- Markov-Bernoulli colouring, 138
- Markov-Bernoulli marking, 138, 139
- Markov-Bernoulli recording, 134, 135, 136, 137, 138, 139
- Markov-Bernoulli thinning, 140

Markov-Lindley process, 186
Markov-Poisson process, 28, 29, 32, 105
matrix
- expectation matrices, 175
- Laplace transform matrices, 176
- matrix-algebra/algebraic, 156
- matrix-geometric, 108
- matrix-measure, 88
- probability matrix/matrices, 17, 76, 88, 160, 185, 186, 196
- sub-stochastic matrix, 178
- transition matrix, 4, 67, 71, 99
- transition probability matrix, 17, 75, 88, 160, 185, 186, 196
- transition rate matrix, 160

maximum
- maximum function, 49, 72, 144
- maximum functional, 49, 72, 144
- maximum number, 15, 44
- maximum output, 156, 191
- maximum rate, 155, 190

MCPP (see Markov-modulated compound Poisson)
measure
- atomic measure, 175
- Borel measure, 63
- conditional distribution measure, 36
- distribution measure, 10, 17, 24, 31, 33, 36, 50, 52, 54, 55, 64, 93, 144, 176, 180
- initial measure, 10, 65
- finite measure, 57, 175
- ladder measure, 143
- Lebesgue measure, 55
- Lévy measure, 37, 50, 56, 57, 59
- Markov renewal measure, 51, 52, 76, 145
- matrix-measure, 88
- probability measure, 58, 63, 72, 73, 74, 82, 83, 99, 109, 111, 125, 126, 127, 129, 132, 136, 144, 180
- renewal measure, 51, 52, 76, 86, 145
- stationary measure, 99, 144
- transition distribution measure, 10, 17, 24, 31, 33, 50, 52, 54, 64, 93, 144
- transition measure, 33, 62, 81, 85, 93, 94, 100, 127, 136, 144, 151
- transition probability measure, 72, 73, 74, 99, 109, 111, 125, 126, 127, 129, 132, 136, 144

MMPP (see Markov-modulated)
modulating Markov chain, 8, 72, 94, 141
moments
- finite moments, 92
- mean, 9, 13, 27, 29, 31, 54, 63, 69, 72, 81, 95, 105, 106, 124, 148, 174, 185, 195
- moments of the number of counts, 103, 114, 119, 121
- second order moments, 121

MRP (see Markov renewal process)
MRW (see Markov random walk)
multivariate, 104, 113, 114, 133

net input
  net input process, 160, 166, 189, 191, 194, 197
  net input rate, 156, 156, 157, 157, 194, 194, 201, 201
network
  ATM network, 108, 130
  communication(s) network, 108
  communication systems, 155, 157, 159, 161, 163, 165, 167, 169, 171, 173, 175, 177, 179, 181, 183, 185, 187, 190, 191
  Markovian network, 111
  queueing network, 103, 112, 134
  storage network, 158

operator
  linear operator, 114, 177
  operator notation, 19
  operator semigroup, 177
  shift operator, 163, 181
output
  output channel, 156, 158, 191, 199
  output process, 108, 160
  output rate, 156, 191
partition
  block-partition, 198
  finite partition, 139
period
  idle period, 143, 153
  busy-period, 145, 146, 153, 159, 170, 171, 172, 173, 174, 178, 179, 180, 186, 200, 201
phase-type
  compound phase-type arrival process (CPH), 112
  phase-type arrivals, 107, 108, 114
  phase-type arrival process, 117
  phase-type distribution, 107, 117
  phase-type Markov renewal customers, 112
  phase-type Markov renewal process (CPH-MRP), 117
Poisson process
  compound Poisson process, 8, 32, 34, 55, 106, 107, 134, 139, 159, 165, 177
  Markov-compound Poisson process, 8, 32, 34, 159
  Markov-modulated Poisson process, 8, 28, 105
  Markov-Poisson process, 28, 29, 32, 105
  (standard) Poisson process, 29
  switched Poisson process, 131
preemptive priority, 191, 199
probability
  probability density, 116, 117
  probability distribution, 73, 83
  probability measure, 58, 63, 72, 73, 74, 82, 83, 99, 109, 111, 125, 126, 127, 129, 132, 136, 144, 180
process (other)
  branching process, 8
  counting process, 9, 86, 137, 138, 139
  Cox process, 28, 106
  death process, 155, 157
  Lévy process, 5, 8, 24, 36
  loss process, 140
  thinned process, 139
  overflow process, 136
  point process, 104, 109, 117, 118
  reward process, 185
  Rudemo's process, 107
property
  additive property, 53, 62, 103, 124, 130, 133
  class property, 68
  lack of memory property, 30, 40, 103, 114, 115, 116, 143, 152
  Markov property, 12, 12, 42, 42, 48, 48, 66, 66, 68, 68, 75, 75, 79, 79, 130, 130
  Markov-additive property, 52, 103, 124, 133

Markov renewal property, 31, 33, 44, 47
regenerative property, 52, 84
renewal property, 31, 33, 44, 47
semigroup property, 115
semiregenerative property, 52
strong Markov property, 12, 66, 68, 75

quasi-Markov
quasi-Markov chain, 39, 40, 42, 48
quasi-Markov process, 40
queue
M/G/1 queue, 141, 180, 187
M/M/1 queue, 136, 137, 141, 143, 151, 152, 154, 190, 200
queue length, 190
queueing
queueing model, 8, 106, 141
queueing network, 103, 103, 112, 112, 134, 134
queueing system, 8, 24, 100, 103, 104, 112, 133, 134, 141, 143, 144, 145, 147, 149, 151, 153, 155, 190, 194
single server queueing system, 100, 141, 143, 145, 147, 149, 151, 153

random
random clock, 165, 170
random index, 98
random experiment, 65
random process, 1, 62
random sequence, 127
random time transformation, 103, 132, 133
random transformation, 128
random variable, 2, 3, 12, 14, 15, 24, 29, 33, 41, 42, 50, 70, 94, 95, 142, 143, 144, 145, 150, 153, 155, 173, 185, 196, 203
random walk, 1, 8, 24, 41, 49, 61, 62, 63, 64, 65, 66, 67, 68, 69, 71, 72, 73, 75, 77, 78, 79, 81, 83, 84, 85, 87, 89, 91, 93, 95, 97, 99, 101, 141, 142, 150, 189, 192
range
observed range, 12, 15, 39, 44, 63, 66, 67
range of, 3, 12, 39, 40, 42, 43, 44, 46, 52, 54, 55, 57, 67
rate
arrival rate, 123, 123, 137, 137, 151, 151, 177, 177, 180, 180, 191, 191, 200, 200
birth rate, 157
constant rate, 156, 190
consumption rate, 156
death rate, 157
demand rate, 157, 159, 160, 174, 184
input rate, 155, 156, 157, 159, 191, 194, 201
intensity rate, 106
jump rate, 174
limit rate, 188
maximum rate, 155, 190
net input rate, 155, 156, 157, 194, 201
output rate, 156, 191
production rate, 156
rate matrix/matrices, 107, 131, 160
rate of occurrence, 8, 29
rejection rate, 180, 188
release rate, 158
service rate, 133, 137, 151, 199, 200
transition rate, 25, 28, 32, 50, 55, 105, 109, 110, 130, 131, 160
recurrence
recurrence and regeneration, 1, 3, 5
recurrence equations, 142
recurrence properties, 66
recurrence time, 3, 43, 44, 47, 48, 49, 58
recurrence time distribution, 44, 47, 58

recurrent
positive recurrent, 124
recurrent event, 1, 2, 9, 39, 42
recurrent phenomena, 2, 3, 5, 9, 40, 43, 48, 58, 63
recurrent phenomenon, 2, 3, 9, 43, 44, 46, 47, 48, 49
recurrent process, 2
recurrent set, 40, 43, 61, 63, 72, 75, 78, 87, 145
regeneration
recurrence and regeneration, 1, 3, 5
regeneration point, 1
regenerative
regenerative phenomena, 3, 4, 5, 8, 39, 40, 41, 42, 43, 49, 56, 57
regenerative phenomenon, 3, 4, 40, 41, 42, 50, 54, 56, 57, 58, 59
regenerative process, 5, 39, 40
regenerative property, 52, 84
regenerative set, 39, 40, 42, 50, 52, 54, 55, 57
regenerative system, 40
strongly regenerative, 43
regularity condition, 85, 93
renewal
renewal component, 11
renewal (counting) process, 1, 3, 5, 7, 8, 9, 10, 11, 12, 13, 16, 21, 23, 39, 42, 43, 62, 63, 66, 67, 84, 95, 107, 108, 117, 142
renewal customer, 112
renewal equation, 9, 19, 20, 21, 22, 61, 63, 85
renewal function, 19, 21
renewal function, 19, 21
renewal sequence, 128
renewal theorem, 23, 82
renewal theory, 9, 13, 16, 19, 21, 39, 43, 57, 146
representation, 113, 114, 117, 118
risk, 8, 62, 101
sample
sample function, 5, 17, 19, 112, 115, 163, 191, 202
sample path, 17, 52, 54, 55, 161, 162, 163, 164, 167, 168, 169, 170, 197
secondary
secondary arrival process, 134
secondary Markov chain, 141, 142
secondary process, 103, 134, 135, 136, 139
secondary recording, 103, 134, 135, 136, 139
semi-Markov process, 9, 22, 23, 117
semirecurrence
semirecurrence time, 43, 44, 47, 48, 49, 58
semirecurrence time distribution, 44, 47, 58
semirecurrent
semirecurrent phenomena, 9, 40, 43, 48, 58, 63
semirecurrent phenomenon, 9, 43, 44, 46, 47, 48, 49
semirecurrent set, 40, 43, 61, 63, 72, 75, 78, 87, 145
semiregenerative
regenerative phenomena, 3, 4, 5, 8, 39, 40, 41, 42, 43, 49, 56, 57
regenerative phenomenon, 3, 4, 40, 41, 42, 50, 54, 56, 57, 58, 59
regenerative process, 5, 39, 40, 84
regenerative property, 52, 84
regenerative set, 39, 40, 42, 50, 52, 54, 55, 57
sequence
convergent sequence, 15, 72, 84, 86, 98, 124
decreasing/nonincreasing sequence, 72, 91, 167
deterministic sequence, 127
divergent sequence, 80
imbedded sequence, 127
increasing/non-decreasing sequence, 72, 168
sequence of arrival epochs, 108

sequence of imbedded times, 127
sequence of interarrival times, 143
sequence of ladder points, 63, 78, 141, 196
sequence of service times, 143
sequence of random variables, 2, 50, 94, 95
sequence of stochastic process, 49
sequence of trials, 2

sequential analysis, 64

service
service interruption, 112, 113
service rate, 133, 137, 151, 199, 200
service rate mechanism, 141, 143
service time, 141, 142, 143, 151, 156, 180, 188
service time distribution, 143, 188

sojourn
sojourn time, 9, 46, 48
sojourn time distribution, 46, 48

state space
countable state space, 1, 9, 62, 66, 141, 158, 175, 176, 189
finite state space, 54, 59, 66, 68, 94, 107, 192
infinite state space, 192

stationary
stationary distribution, 73, 84, 92, 93, 97, 99, 121, 123, 124, 160, 170, 186, 192, 194
stationary and independent increments, 106
stationary measure, 99, 144
stationary point process, 109
stationary probability, 82, 83, 90
stationary version, 121, 124, 185, 201

stochastic
stochastic behavior, 156
stochastic process, 41, 141, 161, 167, 193
stochastic matrix, 160, 178

stopping
stopping rule, 101
stopping time, 12, 43, 75, 84, 115, 127, 132, 167, 169, 175, 197

storage
storage busy-period, 172, 173, 174, 178
storage level, 171, 179, 180, 182, 185, 189, 190, 192, 194, 205
storage level process, 185
storage model, 62, 155, 157, 159, 161, 163, 165, 167, 169, 171, 173, 175, 177, 179, 181, 183, 185, 187, 189, 190, 191, 193, 195, 197, 199, 201, 203, 205
storage network, 158
storage policy, 158, 189, 190
storage process, 155, 160, 161, 178, 184, 186, 189, 192
storage system, 170, 183, 184, 188

strong Markov
strong Markov process, 30, 116
strong Markov property, 12, 66, 68, 75

subordinator, 5, 8, 39, 40, 42, 50, 52, 54, 55, 105, 106, 115, 126, 132

supremum, 71

system
dynamical system, 1
inventory system, 157

theorem
a Tauberian theorem, 184, 187
central limit theorem, 93, 94, 97, 98, 100, 124, 150

time
occupation time, 40, 53
exit time, 61, 64, 91
hitting time, 13, 33, 48, 66, 94
interarrival time, 103, 105, 107, 114, 116, 117, 141, 143, 151
local time, 180
time-reversal, 142
time-reversed, 61, 63, 72, 73, 74, 79, 87, 88, 90, 99, 100, 143, 144, 148, 196, 197
time transformation, 103, 132, 133

traffic
bursty traffic stream, 130
heavy traffic limit, 100

- traffic generated, 135
- traffic shaping, 109
- traffic stream, 130

transform
- Laplace transform, 175, 176, 177, 178, 188
- Laplace-Stieltjes transform, 180
- Fourier transform, 63
- Pollaczek-Khinchin transform formula, 188

transformation
- deterministic transformation, 125, 128
- linear transformation, 103, 125, 126, 127, 128
- measurable transformation, 125
- patching together, 125, 126, 128, 138
- random time transformation, 103, 132, 133
- superpositioning, 106, 108, 131
- transformation of MAPs, 124, 125, 128, 132
- sums of independent MAPs, 130

transition
- infinitesimal transition, 25
- self-transition, 184
- transition density, 31, 153
- transition distribution, 10, 17, 19, 24, 31, 33, 34, 50, 52, 54, 64, 93, 144
- transition epoch, 107, 123, 128, 191
- transition kernel, 9
- transition matrix, 4, 67, 71, 99
- transition measure, 33, 62, 80, 85, 93, 94, 100, 127, 136, 144, 151
- transition probabilities, 3, 11, 17, 25, 28, 32, 65
- transition probability, 4, 17, 72, 73, 74, 75, 76, 88, 99, 109, 111, 116, 117, 125, 126, 127, 129, 132, 136, 144, 160, 185, 186, 196
- transition probability matrix, 17, 75, 88, 160, 185, 186, 196
- transition rate, 25, 28, 32, 50, 55, 105, 109, 110, 130, 131, 160
- transition rate matrix, 160

upper bound, 101

variance
- variance component, 192
- variance of the number of counts, 106

waiting time (in queue), 100, 143, 144, 146, 150, 187, 188

weighted average, 147